ROUTLEDGE LIBRARY EDITIONS:
ORGANIZATIONS: THEORY & BEHAVIOUR

MANAGING ORGANIZATIONAL CHANGE

MANAGING ORGANIZATIONAL CHANGE
Human Factors and Automation

FRED FALLIK

Volume 14

Routledge
Taylor & Francis Group
LONDON AND NEW YORK

First published in 1988

This edition first published in 2013
by Routledge
2 Park Square, Milton Park, Abingdon, Oxon, OX14 4RN

Simultaneously published in the USA and Canada
by Routledge
711 Third Avenue, New York, NY 10017

Routledge is an imprint of the Taylor & Francis Group, an informa business

British Library Cataloguing in Publication Data
A catalogue record for this book is available from the British Library

ISBN: 978-0-415-65793-8 (Set)
eISBN: 978-0-203-38369-8 (Set)
ISBN: 978-0-415-82326-5 (Volume 14)
eISBN: 978-0-203-54488-4 (Volume 14)

Publisher's Note
The publisher has gone to great lengths to ensure the quality of this reprint but points out that some imperfections in the original copies may be apparent.

Disclaimer
The publisher has made every effort to trace copyright holders and would welcome correspondence from those they have been unable to trace.

MANAGING ORGANIZATIONAL CHANGE:

Human Factors and Automation

Fred Fallik

Taylor & Francis
Philadelphia • New York • London
1988

USA Taylor & Francis Inc., 242 Cherry St, Philadelphia, PA 19106-1906

UK Taylor & Francis Ltd, 4 John St, London WC1N 2ET

British Library Cataloguing in Publication Data
Fallik, Fred
Managing organizational change: human factors and automation.——(Case studies in ergonomics practice ISSN 0950-9461; v.3)
1. Automation——Social aspects——United States 2. Industrial sociology——United States
I. Title II. Series
306′.36 HD6331.2.U5

ISBN 0-85066-363.6
ISBN 0-85066-362.8 Pbk

Library of Congress Cataloging-in-publication data

Fallik, Fred.
Managing organizational change.

(Case studies in ergonomics practice; v. 3)
Bibliography: p.
Includes index.
1. Office practice—Automation—Psychological aspects—Case studies. I. Title. II. Series.
HF5548.F274 1988 658.3 87-18027

ISBN 0-85066-363.6
ISBN 0-85066-362.8 Pbk

Cover design by Russell Beach
Typeset by Chapterhouse, The Cloisters, Formby, Liverpool L37 3PX
Printed and Bound in Great Britain by
Redwood Burn Limited, Trowbridge, Wiltshire

Contents

Dedications

To my parents, Anne and Solomon.

To my in-laws, Lillian and Alex.

To Jim Lantonio, the human in the human factors.

To the employees of the Automated Collection System.

Preface

The relationship between people and work has always been a love–hate affair. At one extreme, individuals have often viewed work as a necessary evil, to be accomplished with the minimum effort and maximum results. At the other extreme, work has become an end in itself rather than a means to an end. Interposed between human efforts and the results of those efforts is the technological system. This is the tool through which the work is accomplished. The advance of technologically mediated work has accelerated exponentially over the past century. Buckingham (1961) pointed out that "There has been more technological knowledge gained in the past fifty years than in all previous history and the growth is accelerating".

What Buckingham saw from his vantage point in the early 1960s was only the beginning of the revolution brought about by automation. However, the availability of any new technology can have consequences for either degrading or improving the human condition. Like any tool, technology is value neutral. The application of new work technologies tends to reflect the underlying values of those who apply it.

Office automation, while a relatively recent phenomenon, has had a profound effect on the way business is accomplished. John Howard, the president of Wordpro, Inc., points out that office automation has, in its brief history, progressed through three stages of evolution (Howard, 1983). The first stage, identified as the "accounting era", involved large, mainframe computer facilities, which were controlled by centralized authority, performed specialized functions, and operated almost exclusively in a batch mode. This stage was followed by the "operational era", typified by the advent of the minicomputer, local control, specific function related equipment, and on-line availability. This era spawned the first word processors and freed local units from dependence on a centralized data processing division. The third stage, termed the "informational era" places emphasis on the integration of systems, multifunctional devices, local or even individual control, and the use of microcomputers throughout all levels of the organizational hierarchy.

Two outstanding characteristics of this evolution are its rapidity and its relationship with technological advances. It would probably be a fair statement that no other revolution in the history of mankind has had so profound an effect on a wide variety of industries, in so short a period. The advent of inexpensive, user-accessible computer technology has radically and permanently altered the way factory and office tasks are accomplished.

Few would argue with the position that office automation will continue to expand within the foreseeable future. This tidal wave of organizations rushing to automate appears to be partly based upon the ill-founded notion that automation *per se* is a panacea for organizational ills of many dimensions. It is not. Stettner, as early as 1966, identified several trends that appear to occur with the increasing automation of work:

Many completely unskilled routine jobs are disappearing.
There is an increasing need for workers with skills at a higher level and wider range.
The content of semi-skilled jobs is changing, with less emphasis on manual dexterity and more emphasis on mental abilities, perception, knowledge, and responsibility.
There is a blurring of the distinction between manual and intellectual labor.
Automation is taking over tasks which are less pleasant, more strenuous, repetitive or routine. Workers, as a result, have greater mobility and use of judgment in their work.

The issue is not whether or not automation will continue to expand but the effectiveness of automation in improving productivity *and* the quality of work life. If the sole purpose of automation is to improve productivity without regard to its effect on the quality of work life, the Luddite reaction of 1811 to destroy the new work technology will become commonplace in our time. As some present day managers are becoming painfully aware, there are many effective and costly ways to destroy technology than simple physical destruction.

As many have termed it, this "second industrial revolution" powered by inexpensive computer technology is a double edged sword. On the one hand, the potential of automation to reduce or eliminate onerous, difficult, or dangerous tasks, create new jobs, improve productivity, and promote creativity is almost limitless. On the other hand, automation can result in such consequences as job burnout, morale and motivation problems, high staff turnover, loss of capital investment and enormous productivity drops. One of the most significant factors influencing the long term success or failure of automation is the manner in which automation is designed, implemented and supported in the organization. The importance of human factors considerations in these processes accounts for a great deal of the variance in success or failure of the technology involved. While technology is relatively easy to identify, catalog, cost out, and measure, the human resources necessary to implement, operate, and maintain new work technologies are much less concrete. Unlike technology, human resources are not the property of the organization. Since they are more difficult to deal with in almost every way, there is a widespread propensity to ignore or play down their importance.

It is only recently that recognition of individual and group contributions to technological effectiveness has taken on a significant momentum. Quality circles, participatory decision making, job enrichment, socio-technical systems design, corporate day-care and health centers, etc. are only a few of the programs being implemented in industries that, a few years ago, would have not even entertained such notions. While one would like to ascribe this movement to increasing humanism on the part of managers and executives, in reality these changes in worklife are as much the result of increased national and international competition as they are the result of cultural pressures and changes.

The innate tendency of any organization is to survive. The primary means of survival is profit. The term 'profit', of course, means different things in different contexts and organizations. While sales and money are the most common criteria, budgets, staff size, and facilities, are among many other indicators of success. To the extent that individuals and groups contribute to the profit of a division or organization, they are valued and rewarded. Unlike technology, however, the contribution of people is often indirect and collective rather than direct. With the successful models of the computer industry and Japanese management style as evidence, it has become apparent to traditional management that such indirect factors as commitment, loyalty, motivation, and creativity are requisites for survival in a turbulent market and not just desirable attributes. The need to attract and retain a skilled and motivated workforce has become, increasingly, an economic necessity as well as being socially desirable. Executives and managers across a broad range of industries and services are beginning to accept the fact that investments in the human resources of an organization often results in economic benefits equal to or exceeding those of automation. The emergence of human factors, management sciences, and industrial psychology are testament to the increased value of the human element in the total economic cycle.

The purpose, scope and audience of this book

The purpose of this text is to describe and interpret the experiences of a large agency of the United States government prior to, during, and following the introduction of automation into a major functional division. The focus of this description is on the 'human factors' aspects relating to automation. To a large degree, these experiences, 'lessons learned' if you will, may be generalized to many large public and private organizations. This book will attempt to describe how people experience the introduction of office automation. From this 'users' point of view we frequently get a very different picture of automation than we would get from that of the project administrator's.

The origins of this program to introduce automation into a major function within the Internal Revenue Service will be described in some detail later in this text. However, it should be noted here that the study design and data collection efforts described here were typically less than ideal. Indeed, efforts to measure and assure the inclusion of human factors in the automation were restricted by the realities of time, legal restrictions, client motivation, and staff availability. Like most field studies, it lacks the controls that are the hallmark of a true experiment. It is, at best, a quasi-experiment. The information upon which the text is based comes primarily from questionnaires, observation, group and individual interviews, literature reviews and author's own experience as a management consultant. In other words, both quantitative and qualitative data form the bases upon which judgments, implications and interpretations are made.

Some of the material upon which this text is based was drawn from documents and reports developed for IRS internal use. Such items, originally developed at government expense, are not subject to copyright laws and are to be considered in the public domain. When these products are presented in this text they will be identified as to their source.

This book addresses two distinct audiences. The first is the executive or management consultant who has some responsibility for the introduction of new technology. Hopefully, the experiences, guidelines and recommendations that comprise this text will serve as human factors guideposts for their own efforts to introduce automation. The second audience at which this book is directed is the academic. The study design, the quantitative analysis of data, the issues that arose in the course of the study, as well as the conclusions that were reached, should provide a wealth of insights regarding the more theoretical bases of human and organizational behavior under change conditions.

In keeping with the goals of this book, the chapters have been divided into three broad sections. Section I presents the prevailing attitudinal and contextual background to automation and human factors. The chapters in Section II describe the study of the automation effort and the results that ensued. Section III provides a summary of "Lessons Learned" both as a direct result of the automation effort and a summary of relevant but indirect evidence stemming from the literature and the author's experience with organizational change.

Fred Fallik
Washington, DC
March 1987

Disclaimer

The positions, conclusions, and recommendations presented within this book represent the personal opinion or judgment of the author in his capacity as a private citizen only, and do not reflect the position or official policy of the Internal Revenue Service, the Department of Treasury, or any other branch of the United States Government.

The publication of this text was subject to the review and approval of the Office of Public Affairs of the Internal Revenue Service, the Department of the Treasury of the United States of America.

Every effort has been made to describe what occurred in a accurate fashion. Any inaccuracies can be attributed solely to the author.

Acknowledgements

It is a rare experience when one person can take credit for writing an entire text. This work is no exception. The original data and observations resulted from the efforts of many people. In particular, the staff of the Human Resources Technology Staff were instrumental in the collection of relevant information in addition to the development of human factors programs and recommendations.

Several people had a direct impact on the development of the text material itself. In no particular order, these are:

Larz Pearson
Jim Lantonio
Mary Kincaid

Section I

BACKGROUND

Chapter 1
Automation, Beliefs and Values

Introduction

The purpose of this chapter is to set forth some of the attitudes and assumptions which the author believes influenced the introduction of automation into the IRS. These beliefs form an important part of the background to understanding the processes involved in establishing the Automated Collection Service (ACS).

The introduction of automation into business and industry has been viewed as primarily a process involving equipment technology, financial resources, training and strategic planning. In other words, a problem of organizational administration. The successful introduction of automation, although heavily influenced by such traditional technical and administrative practices, is also directly affected by the reaction of organizational members to that introduction. Zuboff (1982) points out some consequences of poor co-ordination and attention to human factors in automation:

> What happens when a group of people are suddenly made to feel powerless and undervalued because of the introduction of computer technology to their jobs? First there are obvious problems of morale and motivation. If employees depend upon the computer system in order to accomplish their work, but the system is experienced as an enemy, then the quality of the work must suffer.

Attitudes, values and beliefs are, of course, built up over a period of years as the result of experiences. Certainly, the successful introduction of new work systems depends upon the restructuring of task and work flow. Of equal, if not greater importance, is the predisposition of organizational members to that change.

1. *Management beliefs and values*

Despite the expense and difficulty required to implement automation in offices and factories, the belief exists among managers and executives that the long-term benefits of competitive advantage, communication, file maintenance and productivity are well worth the short-term expense. Automation, in other words, is ultimately justified in terms of productivity and reduced operating costs, not directly in terms of quality of work life.

While lip service is often paid to the potential for the improved quality of work life that might result from automation, it is difficult to quantify such benefits in terms of dollars and cents. Job satisfaction, morale, creativity, motivation, even such aspects as employee turnover and union grievances, are generally considered 'soft' measures of performance. For the typical executive, these potential benefits are considered to the extent they impact on the 'hard' measures of units, time, errors and money expended. Human performance is subject to differing and changing motivations. Not only are these qualities difficult to measure, they are unpredictable. This situation is undesirable, especially at lower levels in the organizational hierarchy. To the extent that humans can be replaced by machines, the entire system will work with fewer errors, less cost and more effectiveness.

If human issues arise at all, it is in terms of a reduction in human errors rather than increases in creativity, reduced boredom and fatigue. In some measure, this attitude is a consequence of the position and history of the executive or manager. Such high-level positions are rarely boring, unchallenging, machine-paced, or repetitious. A great deal of commitment to the organization is present. It is easy for individuals at this level in the organization to ignore the psychological and social consequences of degraded jobs when one is not directly involved in them. Few executives spend a great deal of time on the shopfloor. Far fewer would be willing to spend a day performing entry-level tasks.

An additional belief that may be held by executives has been termed 'paying your dues'. In rising through the organizational hierarchy, an individual frequently has to hold jobs which are dull, repetitive and unchallenging. Having performed these roles in the past promotes the belief that the experiences were valuable to future success. In some ways, the performance of undesirable jobs is considered a necessary facet of management positions. Positions requiring challenge, freedom, responsibility, etc. are considered the prerogatives of management. Pressure to incorporate these aspects in lower level jobs is taken as evidence of the evils resulting from a liberalized society in which people 'have it too easy and don't appreciate the value of an honest day's work for a fair day's pay'.

Such attitudes, frequently unspoken by managers and executives, are certainly stumbling blocks to humanizing the workplace. Since it is not socially acceptable to express such beliefs in public, many managers and executives maintain a pseudo-humanistic or neutral posture toward rank and file positions. Arguments centering on the critical importance of a skilled and motivated workforce do little to alter such embedded attitudes. What does have an effect is proof that investment in human factors is an extremely cost-effective approach to improving productivity. Ideally, we should be able to quantify the debits and credits associated with human performance at different levels of effectiveness. The issues in developing such measures are complex and often value-laden. Although tedious, it is possible to develop indices representing the cost of labor. The profits resulting from the individual efforts at work are even more nebulous. Since a product or service results from the combined efforts of numerous individuals, it is almost impossible to ascribe to an individual the net contribution of his or her effort to the totality. Indeed, many human aspects such as creativity are not easily measurable. In contrast, technology is much simpler to quantify in terms of costs and benefits. In this sense, it makes more sense to justify the investment in hardware and software than it does in the people who use it.

For the foreseeable future it is unrealistic to expect the situation to change to one in which the improvement of human factors in the workplace becomes an end rather than a means to improve productivity. With the acknowledgment of the many exceptions that exist, unless managerial attitudes change, the competition for labor becomes acute, or acceptable techniques are developed which quantify the worth/contribution of human performance, those interested in promoting human factors will be constantly playing the role of the weak sister.

2. Employee beliefs and values

The reaction of employees to the introduction of automation appears to fall within one of three broad categories. We would estimate from our experience that about 10–15% of employees are enthusiastic and supportive of the technology. Similarly, about 10–15% are very resistant to its introduction. The vast majority of employees, about 70–80%, are ambivalent and generally take a wait-and-see attitude. The general relationship between degree of change and employee reaction is illustrated in Figure 1–1.

These percentages vary from organization to organization and depend on several factors, including prior experience with electronic technology, employee age, length of employment, nature of the industry, and degree of task frustration experienced under the manual system of work. Younger workers who have grown up with home computers and video games, who have not vested a great deal of skill or knowledge in learning manual task performance (i.e., those who don't have to unlearn acquired abilities), or who viewed manual performance as a frustrating, time-consuming experience, tend to be positive about the introduction of automated systems. In contrast, older employees who have vested years of experience in learning the intricacies of the manual system, who doubt their competence with electronic systems, or who see themselves as having achieved successful performance with the manual system, are much less likely to greet the introduction of new technologies with approval.

Figure 1–1. The relationship between change and employee reaction.

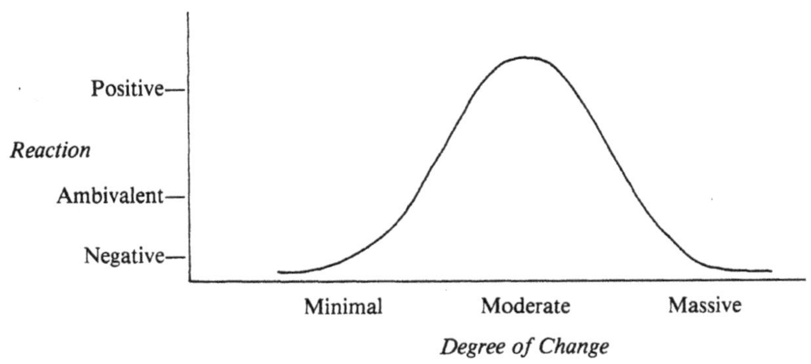

Little in the way of literature exists on capitalizing on the positive attitudes of those employees who see automation as a positive force. The literature is rife with information regarding resistance to change. This imbalance partly results from the common human reaction to dealing with problems before dealing with opportunities and partly from the more vocal and behavioral nature of those who are resistant to the implementation of new work systems. If people are positive about forthcoming changes, the organization will, typically, take a 'leave well enough alone' attitude. Resistance to change, on the other hand, can become manifest in a number of undesirable behaviors: quitting, early retirement, union grievances, work slow-downs, sabotage, etc.

Historically, the implementation of new work technologies has been broadly resisted rather than greeted due to the consequence of the new technologies. Employees, correctly for the most part, have viewed work methodology changes as attempts to promote efficiency by increasing productivity and reducing costs. While the organization has stood to benefit from these changes, the results for the employee have often been lay-offs or increased performance for static pay. Improvements in the quality of working life, if they have occurred at all, have frequently been incidental to changes in the work technology, structure or process, and not a direct effort on the part of the organization. Frederick Taylor's (1947) school of 'scientific management' was one of the few attempts to link increased productivity resulting from improvements in work methods to positive consequences for the employee. Specific attention aimed at improving the quality of work and working conditions is a very recent phenomenon. Marguiles and Colfresh (1982) put this approach in perspective:

> Traditionally, organizations have not balanced their concerns for technological and human subsystem change. The result has been "misfit" which has the apparent consequences of lowered morale, decreased productivity, lower quality of work, increased conflict, and even sabotage, waste, job stress, and duplication of effort. The final result for many organizations has been increased costs of technological implementation, and often underuse or failure of the new technology.

3. *Behavioral science as a tool of management*

Major changes in the way organizations work may be simultaneously both threatening and exciting to the individuals involved. The nature and scope of the change, however, are variables which determine whether change is viewed as exciting or threatening. Moderate changes in known or predictable routines frequently generate enthusiasm. The novelty of newness itself tends to motivate many individuals. Radical change during a short span of time, in contrast, is frequently perceived as threatening. In such situations, the threatening aspects of change are the result of increased insecurity.

Many behavioral and social scientists, as well as managers, would prefer a world in which human and organizational behavior were more predictable. The more predictable it is, the more control that can be exercised over outcomes. The more control that can be exerted over a situation, the lower the level of threat that exists. A highly simplified intuitive process for predicting human behavior, common to the thinking of many individuals, is illustrated in Figure 1–2.

Figure 1–2. A subjective model of prediction.

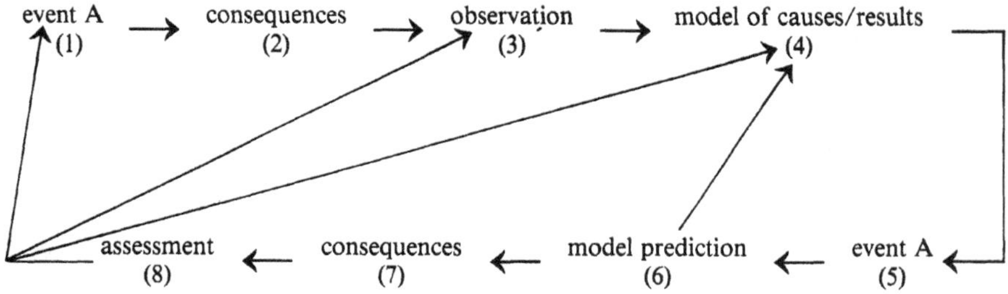

A person observes (3) the relationship between an event (1) and its outcomes (2). Depending on the individual involved, repeated observations of similar situations promote the establishment of a model of human and organizational behavior which lead to conclusions (4) about the relevant factors and causative nature of the relations. Faced with a similar situation at a later time (5), the individual makes a prediction (6)

based upon his or her model. Thus, by using past experience to predict future outcomes, people feel more confident in their ability to control (7) situations. If the prediction is inaccurate (i.e., the outcomes are not what was predicted), the manager may revise the model, define the event (A′) as being significantly different from the model events (A), or may choose to distort or deny his or her perceptions to conform with the expected outcomes. Of the three, probably the least common response is for the manager to alter the models of cause and effect.

In those situations where a major organizational change is planned, the predictability of behavior is degraded. While the established relationships in the individual's existing model may not apply, there is a strong tendency on the part of individuals to continue to apply it in the new situation. The role of the social scientist is to help individuals alter the old models or develop new ones applicable to the new situation.

Such models of human and organizational behavior, as they exist in the minds of employees and managers, should be used as guidelines to predictability rather than precise rules. In a sense, managers would like to be able to emulate the physical sciences by developing laws of human and organizational behavior. Such precision, of course, does not exist when we deal with individual human behavior or that group of individuals we call an organization. Nor is it ever likely to be. A program that works successfully in one organization may be a total failure in another, similar, setting. What motivates one individual may bore another to distraction. The difference between success and failure cannot always be rationally identified. Too often, this lack of rational explanation and predictability leads to frustration and disillusionment. Human behavior, at times, appears to be irrational from the standpoint of other individuals.

The situation, however, is far from bleak. There are many general rules of human behavior whose application tends to improve the predictability of human and organizational behavior. For example, most employees who are criticized in the presence of their co-workers will, almost certainly, become angry and defensive. Just how this anger and defensiveness are expressed will vary from person to person, and even within the same person from situation to situation. From an organizational perspective, lack of attention to employee training practices will usually result in decreased productivity. How this decreased productivity will manifest itself will be a function of the unique situation of that organization and its environment.

In many respects the Internal Revenue Service is a unique organization. The specific conditions that exist within the IRS and its environment are not found in any other organization. The implementation of automation within the Service, in its specifics, cannot be duplicated in any other organization. Yet, by examining the general conditions that tend to promote or detract from the effective implementation of automation, a general model of new work technologies may be derived and generalized to other implementation situations. As is the case with the model of human behavior that individuals develop, the primary goal of this model is to improve the predictability of future events and thus exercise more control over their outcomes.

4. *Models of technology and human factors*

The establishment of the Human Resources Technology (HRT) staff represented the embodiment of a new philosophy regarding the importance of the human element in the IRS. Up to the time of the ACS, the adoption of new technologies within the IRS had been a relatively small-scale, piecemeal process or represented a limited impact.

For example, portable electronic calculators had a widespread impact but required very similar skills to mechanical calculators. The use of mainframe electronic computers, in a different vein, had a tremendous impact, but only with respect to a limited number of computer specialists.

With the advent of the Automated Collection System, the IRS was entering a new era, one in which a great number of people's work structures and management would be dramatically altered. The role of the HRT was to ensure that human factors were not neglected in this transformation.

From a very practical point of view, the HRT staff needed a philosophy with which to approach, organize and 'sell' the need to take into account the mutual relationship between technology and human behavior. Such an organizing principle was found in the Open Systems approach to organizational behavior (Parsons, 1960). With slight modifications to the original concept, the diagram below represents the system elements and their interrelationships (Figure 1-3).

Figure 1-3. An open systems model of automation and human factors.

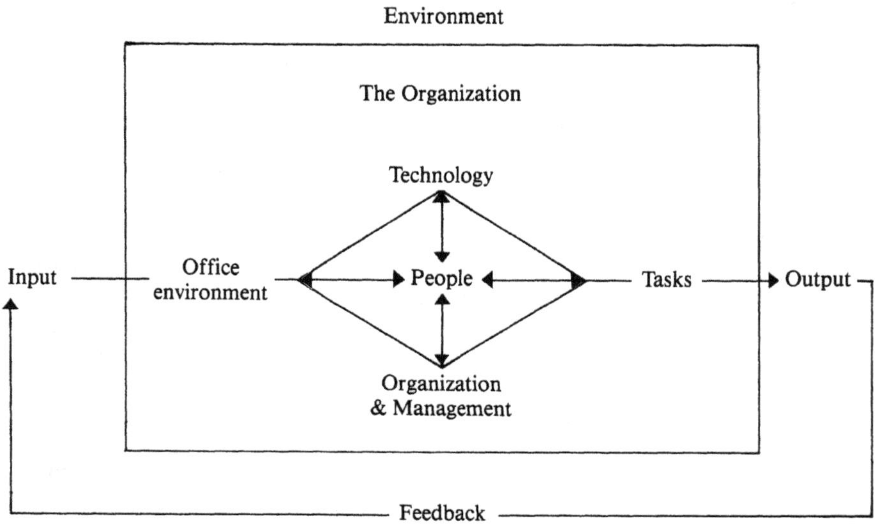

The model is deceptively simple. Depending on the level of analysis of each relationship identified in the model, it provides a systematic framework for looking at the organizational systems that are critical to the successful implementation of a new work system. Putting the model in a matrix format, as illustrated below (Figure 1-4), allows one to focus on these interrelationships and promotes a more detailed approach to planning the factors and consequences of the automation.

Each cell can be analyzed in two directions. For example, one issue that should be raised is the effects of the office environment on the people. Looking at this same issue from the reverse position, the analyst should pose the question, "What effects might people have on the office environment?" The purpose of the model is to focus attention on factors and relationships that might not typically be considered. The inevitable consequence of raising one question is the posing of other, related questions. The matrix provides a starting point. Each cell should be expanded to specify the key elements of that relationship. To use the example cited previously, "What is the

Figure 1–4. A matrix application of the open systems model.

	Focus								
	Input	Envir.	Organ.	Tech.	Office Envir.	People	Tasks	O&M	Output
Input	×								
Environment		×							
Organization			×						
Technology				×					
Office Environment					×				
People						×			
Tasks							×		
O&M								×	
Output									×

Figure 1–5. A matrix of people and office environmental factors.

	People					
Environmental factors	Clerks	Maintenance	Technicians	Managers	Taxpayers	Other
Files						
Break Area						
Workspace						
Lighting						
Noise						
Temperature						
Access						
Storage						
Security						
Visual App.						
Other						

relationship between people and the work environment?", a further elaboration of the matrix might appear as in Figure 1–5.

Similarly, any element in the matrix can be subdivided into its component systems for additional analysis. For example, the category of 'technicians' might easily be broken down into such subdivisions as new hires, journeyman level (fully competent), and technical managers.

Parts of one subsystem can be compared with major components of another subsystem *ad infinitum*. With regard to the impact of automation, the following components, at a minimum, need to be addressed (with example of each).

Environmental issues

The political environment, transportation, unemployment rates, public image, competition, socio-demographic factors, population life-style, educational level, academic institution, real estate, legal considerations, energy, public utilities, stability (predictability), etc.

Organizational issues

Organizational culture, values, myths, heroes, policies, history, stability, size, diversity, geographical distribution, legal constraints, philosophy, etc.

Technological issues

Equipment, furniture, repair and maintenance, availability, condition, ergonomic considerations, transportability, support conditions, compatability, improvements, replacement, etc.

Office environmental issues

Space available, adaptability, location, personalization, lighting, noise, temperature control, hazards, glare, maintenance, break/rest facilities, bathrooms, storage, fire control, security, air quality, windows, attractiveness, etc.

People issues

Training, promotions, recruitment, knowledge, skills, abilities, motivation, management, job descriptions, evaluation, labor force, retention, union considerations, group interaction, retirement, benefits, willingness to relocate, working hours, work shifts, etc.

Task issues

Structured vs unstructured, complexity, discretion, interface, work flow, interdependencies, bottlenecks, supervision required, training required, staffing, etc.

Organization and management issues

Policies, procedures, philosophy, authority relations, functional/geographical assignments, turnover, career and succession planning, reward and punishment systems, span of control, standards, planning functions, etc.

Input issues

Source reliability, dependence, money, people, materials/supplies, timing, quantity, quality, flexibility, etc.

Output issues

Predictability, quality, quantity, timing, delivery, utility, cost, pricing, profit, goal attainment, etc.

Feedback issues

To whom, amount, degree of detail, topics, breadth, sources, redundancy, frequency, media, by whom, timing, objective/subjective, consequences, cost, etc.

Certainly, this is not intended as an inclusive or exhaustive list. Each issue raised requires additional elaboration. It does, however, provide a starting point and focus for the systematic analysis of the key issues that frequently arise. Like all systems, its success depends ultimately on the amount of thought and effort put into its application.

The application of this system approach within the IRS by the HRT staff proved to be both an educational experience as well as a diagnostic instrument. While the open systems model forces consideration of the interrelationships among a wide variety of organizational factors, it became apparent that most managers did not consider all systems to be of equal importance. In a similar sense, the interactions among system components were typically ignored or trivialized. In the majority of cases, it was the technology which was seen as the driving force behind the need to increase productivity and reduce costs. The other issues (office environment, organization and management, task design and people) were frequently secondary considerations at best. The consideration paid to these secondary factors and relationships was the result of the desire to accomplish the goals of the automation effort. In these cases, the use of the systems model helped to refocus attention on all system components.

Except in very rare instances, the philosophy underlying the open systems model failed to win wide acceptance within the IRS. While the reasons for this lack of acceptance were numerous, there are several that tend to stand out. Primary among these was the relative inertia among people who were accustomed to dealing with the bureaucracy in a specified manner over a period of years. Patterns of behavior that were successful in the past were very difficult to give up. Second, use of the open systems approach required a great deal of time and effort. The increase in effort required was not always appreciated even if the final product was superior. Finally, the need for cross-functional co-ordination became very obvious in the open systems analysis. The need to co-ordinate sometimes conflicted with the philosophy of vested interests and territoriality that existed among some participants.

5. *Patterns in the implementation of automation*

The automation of a function or activity appears to follow a fairly consistent and predictable pattern. Our experience with the Automated Collection System and similar efforts in other agencies led to the identification of a common pattern among these interventions. This model is displayed in Figure 1-6.

The model depicts productivity levels along the vertical axis. Productivity is defined as the relationship between profits and costs. Inherent here are the notions of organizational mission, effectiveness and efficiency. The horizontal axis represents the passage of time. As long as productivity levels remain within an acceptable range, few attempts will be considered to alter organizational structures or processes. Even when productivity drops below a satisfactory level, organizational inertia will act to resist any attempts at major revision. Activities within this 'below satisfactory' area are typically of the band aid approach; minor adjustments or remedies which deal with symptoms rather than fundamental issues. Over a period of time, performance which stabilizes within this range may tend to become accepted as the norm. Previously higher levels of productivity are attributed to better times and more favorable conditions that existed in the past.

If productivity levels continue to drop below the zone of inertia or are seen as less inevitably falling below a critical threshold, major changes in the organization are

Figure 1-6. A model of productivity levels before, during and after the introduction of automation into the workplace. Productivity is defined solely in terms of cost and profit.

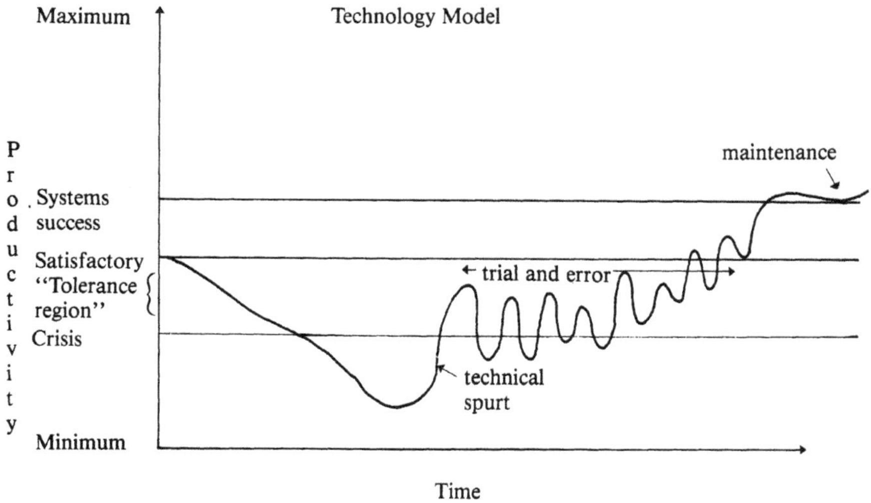

Figure 1-7. A model of productivity levels before, during and after the introduction of automation in the workplace. Productivity here is defined not just by profits and costs, but also takes into account the human factors benefits to the system.

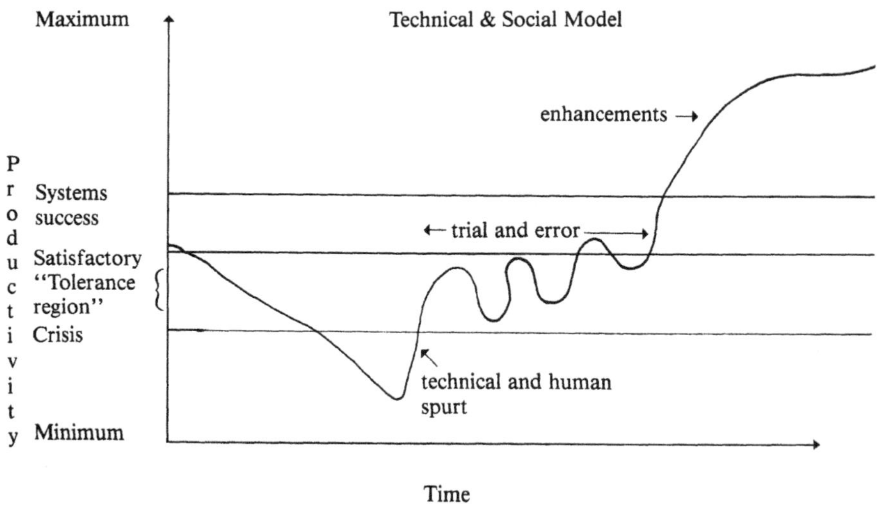

proposed. These proposed changes may take many forms including reorganization, divestment of major activities, or changes in the organizational mission or markets. Appropriate within the current context was the decision to introduce automation as a remedy for depressed levels of productivity.

The introduction of automated technology (new equipment and systems) has a dramatic and immediate impact on productivity. The advent of electronic word processing, for example, provided a considerable and immediate improvement compared to manual typing systems. Since new systems are rarely introduced without some birth pains, the introduction of new technologies requires a period of trial and error. It is during this period that new applications, unanticipated results and system limitations are tested. Some testing results in sharp increases, some in decreases, but the overall trend is in the direction of increased productivity. In general, the emphasis here is on adapting the new technology to the tasks to be performed. Over the course of time, these technical problems are solved with the result that the system is defined as a success. Having thus achieved its projected levels, system maintenance needs to become a primary focus until such time as the entire cycle begins again when productivity decreases below a minimally acceptable critical level.

This technological model of systems innovation provides a rationale for the activities associated with major organizational changes. It depicts the fairly common process on which planners predict their expectations. An alternative model, based on the inclusion of human factors considerations in the technological change, is illustrated in Figure 1–7.

The human factors model is, in some respects, a parallel to the technological model. The same initial forces may be assumed to be responsible for establishing the need for major organizational innovation. The initial technical spurt may be seen to occur but exhibits a sharper slope since system end-users have been involved in the planning, design and implementation of the new systems. The familiarity and mental 'de-bugging' that this approach promotes allows for a much more thorough pre-testing and anticipation. The involvement of users prior to and during the system implementation also results in a decrease in user resistance to change and a heightened motivational level since the success of the new system becomes an individual matter rather than an organizational productivity program (Katzell and Yankelovitch, 1975; Strauss, 1977).

A human factors process, similar to the trial and error system debugging, may also occur with respect to the individual element. Rather than technological variations and adaptations, the human element tends to revolve around new role relationships, work interactions, peer relationship, work expectations, etc. While some of these are planned (e.g., new job descriptions), many evolve as a result of the ambiguity inherent in the implementation of the new technology itself. Like the interaction of the new technology, some of these human factors changes result in spurts of productivity and motivation, and some act to decrease productivity. Overall, the trend should be upwards as systems become stabilized. These human factors activities should supplement and support the technological changes such that the entire productivity curve increases rapidly compared with the technological model.

The peak of the human factors model, unlike the technological model, exceeds the projected level of system success. Since typical systems planners find it difficult to quantify the productivity consequences resulting from the application of human factors in the system implementation, their projected level of productivity is based solely on the technical specifications, assuming employee acceptance. The human factors model is predicated on a higher level of employee involvement, creativity and motivation. In almost all cases, this will result in a higher level of productivity than anticipated and a sharp decrease in time spans.

The introduction of new technologies and major work system changes frequently result in a high level of employee motivation similar in result to the human factors model. This high level of motivation and involvement may however, be an artifact of the change itself rather than an increase in organizational commitment. Individuals approach such major organizational changes with a certain amount of trepidation and resistance but also with a positive attitude in the belief that things will improve. The effect of novelty can easily be mistaken for increased employee dedication. During the implementation period, people will exert an effort and dedication atypical of their pre-change behavior. Based on the technological model, this pseudo-increase rapidly decreases once the installation is completed and the system designated a success. Using the human factors model, the establishment and maintenance of employee participation becomes an on-going pattern. It is this involvement that promotes full system utilization and adaptation to the goals of the organization.

The transferability of the models, conclusions and results described in this book depends on the degree to which organizational and personnel characteristics coincide with those of the people and functions described here. It is my belief that the vast majority of organizations and individuals have much more in common than they have differentiating them. In chapter 2, a brief description of the Internal Revenue Service and the Collection Division, and in chapter 5 of the Collection Office Function/Automated Collection System are presented. In these descriptions, those organizational and structural factors which influenced the implementation of automation will be described.

Chapter 2
The Background to Automation

1. The Internal Revenue Service

The IRS is the largest bureau within the United States Department of the Treasury. The mission of the IRS is to ensure fair and consistent application of the tax laws and to promote voluntary compliance with these laws.

The implementation of automation must be considered within the context of the existing organization. Since no two organizations are identical, the processes and outcomes of implementation will, in specifics, be unique to each organization. Among other factors, organizational structure, size, geography, history, personnel and culture influence the course of the implementation.

The introduction of automation within the IRS, for these reasons, must be thought of as unique. First, the IRS is a large federal agency employing about 85000 people. As a federal agency, the IRS must conform to congressional and judicial rules affecting personnel practices, procurement policies, funding appropriations, etc. that are not found in the private sector. As a large, geographically dispersed organization of career civil servants, the amount of effort that is required to overcome the inherent inertia of almost any structural change requires more effort and time to accomplish than it would in a smaller agency.

A second important distinction regarding the IRS as an organization concerns the inherent stability of its existence. As the saying goes, "Death and taxes are the two inevitable facts of life". Few organizations have the stability enjoyed by the Internal Revenue Service. This organizational stability has its roots in the Congressional Act of 1862 which established the Office of the Commissioner of Revenue. In addition, the fact that the IRS has never had a major across-the-board firing or 'reduction in force' (RIF) to date, coupled with civil service job security promotes an attitude of employment and procedural stability.

Organizational culture

Dealing as it does with money and laws, the Internal Revenue Service tends to be somewhat more conservative in its approach to change than many other federal and private sector organizations. The reasons for this are tied directly to its mission and external environment. An atmosphere of conservatism permeates the majority of IRS activities since rules, regulations, precedents and procedures need to exist to cover the extremely complex series of laws and rulings which govern tax assessment and collection. One of the major consequences of the milieu in which the IRS operates is the extraordinary amount of paper, forms, documents, directives, etc. which are part of the everyday existence of the organization and its employees. In large part, the overwhelming nature of this paper avalanche created the impetus for the conversion to automated systems underlying this study.

Many of the rules by which the organization operates are under the direct control of Congressional and judicial authorities which make and interpret tax laws. Since these laws and their interpretation are in a constant state of change, the IRS is continually in a predictable state of flux attempting to apply those changes to the kaleidoscope of tax related activities.

In recent years, the pace and complexity of these changes have increased tremendously. Not only have the number and complexity of tax laws proliferated but the propensity of the public to challenge federal activities has increased manyfold. Recent problems with computer programs, lost tax return information, and delayed refunds have probably eroded public confidence in the IRS to some extent.

The IRS employee

In terms of its workforce, the IRS, as might be expected, employs mostly those with accounting and auditing skills. Although the general public tends to view the IRS employee as a government accountant, there exist several different related positions encompassed within that broad category. For the following positions, there exist specializations requiring different educational backgrounds and work experiences.

A Revenue Agent is an accountant who examines both individual and business tax returns to determine tax liability.

A Revenue Officer attempts to collect delinquent taxes and returns.

A Specialist Agent, in contrast, conducts investigations of tax evasion to determine whether criminal fraud is involved.

A Tax Auditor examines the returns of individuals and small businesses.

Internal Security Inspectors conduct investigations of allegations concerning current IRS employees and prospective employees, as well as threats against IRS employees.

Taxpayer Service Specialists assist taxpayers by providing information about tax liability and issues or complaints concerning their tax returns.

In addition to these accountant and investigatory positions, there are several large categories of employees who provide support and services to other IRS operations. The three largest of these are personnel specialists, computer specialists, and training and development personnel. While the actual number of individuals employed by the IRS has remained fairly stable over the years, increasing workloads have required a greater and greater effort to simply maintain the status quo. 'Do more with less' has been an operating principle for some time.

The organization structure

Structurally, the IRS maintains a presence throughout the United States and many foreign countries. Most employees tend to divide the organization into two parts, the National Office (NO) and the field. The national headquarters, located in Washington, DC, employs some 4500 people. The national office is responsible for the overall administrative and policy direction of the Service. The headquarters staff also provide the support, development and co-ordination of such service-wide programs as tax law procedures, selection procedures and training activities. The administrative structure of the national office is diagrammed in Figure 2-1.

In geographical terms, the IRS has offices throughout the United States and many

Figure 2-1. The IRS organization chart.

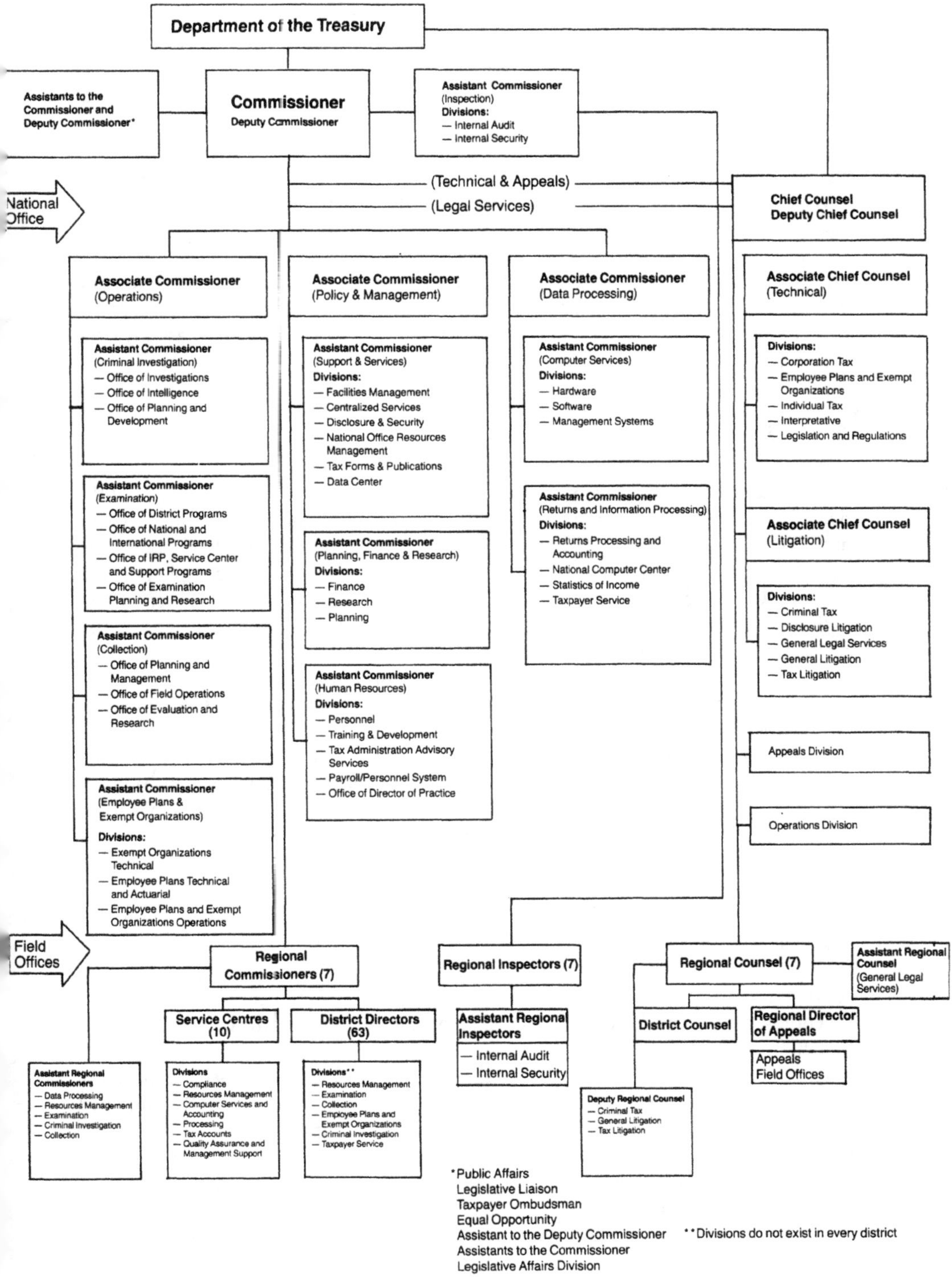

foreign countries. In addition to the National Office, there are 7 Regional Offices and 63 District Offices, 10 Service Centers, the National Computer Center and the IRS Data Center. In comparison with the National Office, the field organizations employ a significantly larger number of people. Each field division serves a defined population area or provides support functions to the National, Regional and District staffs. Figure 2-2 provides an overview of the jurisdictional breakdown of the various divisions.

The IRS, during the course of this study, was headed by Commissioner Roscoe Egger. The Commissioner is the only political appointee in the IRS, all others being career civil servants. Mr Egger was appointed during President Reagan's first term with the approval of the US Senate. Prior to his appointment as head of the IRS, Mr Egger was with the accounting firm of Price Waterhouse. After an initial tour of IRS operations following his appointment, Mr Egger concluded that automation of manual operations should be a top priority. Throughout his career with the IRS, Mr Egger has been a driving force behind the efforts to bring the benefits of automation to the Service.

2. *The Collection Division*

The Collection Division, one of the largest among all those in the IRS, is not what the name implies. Perhaps a more appropriate name would be the 'Delinquent Accounts and Compliance Division'. The purpose of the Collection Division is to collect delinquent taxes, secure delinquent returns, and bring the taxpayer into compliance with the tax laws. A Collection branch office was located in each of the 63 District Offices nationwide. A large branch office numbered from 125 to 130 employees, a medium-size office from 60 to 70 people, and a small office employed about 10–30 people. Altogether, 6–7000 people were employed in Collection Function activities.

The Division was composed of three major functions, the Collection Field Function (CFF), the Special Procedures Function (SPF) and the Collection Office Function (COF). Organizationally, the Field Function was staffed with a Branch Chief, Revenue Officer Group Managers, and usually 12–15 Revenue Officers. The Field Function accepted taxpayer cases which had been processed through the Collection Office Function without reaching a satisfactory resolution. The Field Function's purpose was to make face to face contact with the taxpayer and to physically enforce collection, if necessary, by seizure of taxpayer assets.

The Special Procedure Function was staffed with a Branch Chief and 12–15 Tax Technicians and Reviewers. The purpose of this function was to handle the legal issues for the Collection Division. Frequently these concerned such matters as bankruptcy, court cases, and allegations of bribery and officer misconduct. In practice, this function served as an advisory staff for the Revenue Officers and the Collection Office Function.

The Collection Office Function was headed by a Branch Chief. Under that position were Section Chiefs and Unit Managers. Employee staffing consisted of Revenue Representatives, Tax Account Representatives, and clerical support. The purpose of this function was to collect Taxpayer Delinquent Accounts (called TDAs) for outstanding balances owed on a tax return or by conducting a Taxpayer Delinquent Investigation (TDI) for those individuals or organizations who failed to file a tax return or filed infrequently. Taxpayer Delinquent Accounts were generated from the regional Service Center on a weekly basis. The Service Center, which receives and processes all tax

Figure 2–2. The geographical organization of the IRS within the USA.

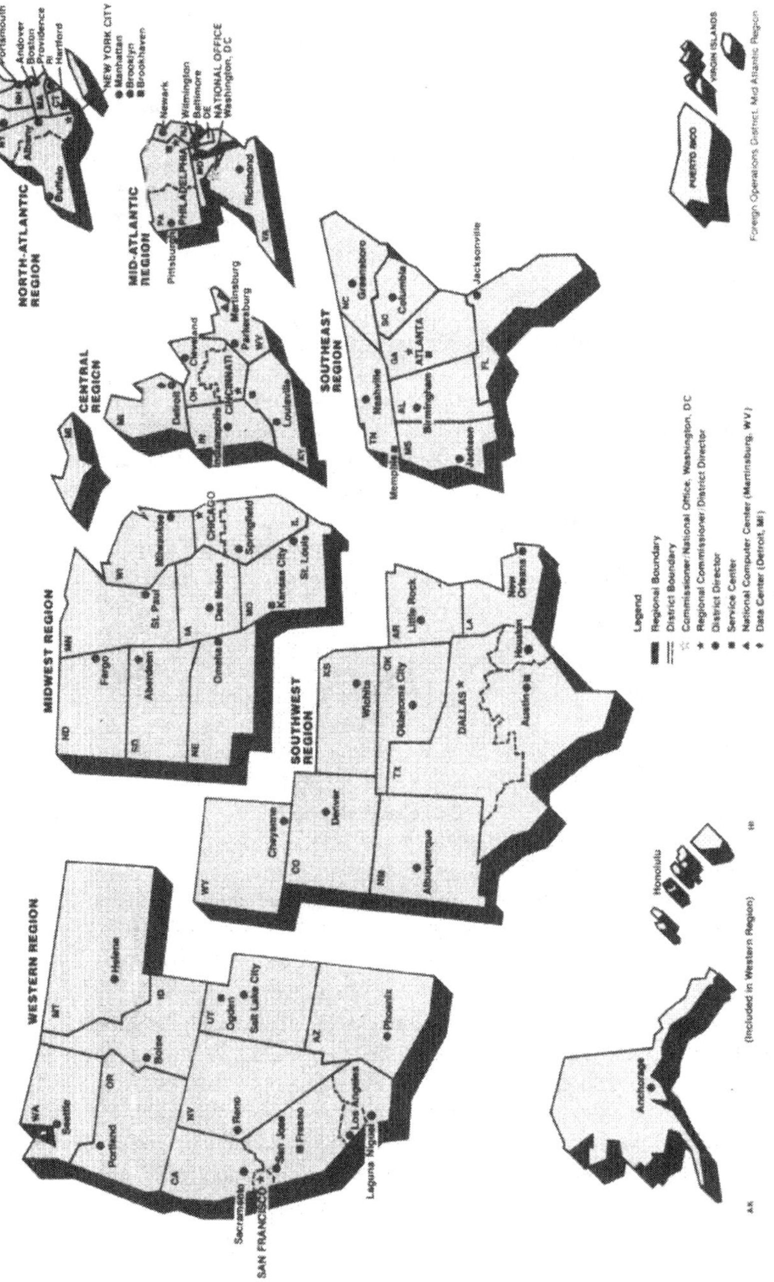

returns within a given geographical area, is required to mail four notices of delinquency to taxpayers having an unresolved tax account. Unresolved tax accounts may occur for various reasons: the taxpayer refuses to pay, the taxpayer claims not to have the assets to pay, or the amount owed is in dispute.

The records of all transactions pertaining to the Taxpayer Delinquent Account were maintained in a paper document format. The 5″ by 8″ index cards used to keep these records contained the name, address, tax period, assessment information, and the cycle of insurance. If a satisfactory arrangement was not obtained after the four written notices were sent by the Service Center, the account was transferred to the Collection Office Function.

When received in the Collection Office Function, the account was initially reviewed by the clerical staff to determine the next actions to be taken based upon the type of tax liability, the amount owed, the history of attempts to collect, etc. Those cases requiring telephone contact were researched through the use of telephone directories. Each COF Branch maintained a complete set of local telephone directories, criss-cross directories (i.e., cross-referenced phone numbers by street address) and telephone assistance operators. Often the telephone directories were out of date. Telephone numbers, if obtainable, were recorded in the case file. Also attached to the case file was a written history sheet in order to allow the Revenue Representative to document case actions and progress.

Following actions to obtain taxpayer phone numbers, the case was sent to the Contact Function. In this Function, telephone calls were made to contact the taxpayer. If the attempt to contact the taxpayer was unsuccessful, the file was returned to the clerical unit. If time permitted, the case was recycled and a second attempt to contact the taxpayer was made. Frequently, the overwhelming inventory of backlogged cases prohibited such a follow-up attempt.

Based on the actions identified on the case history sheet, within 15 days of failure to resolve the outstanding tax debt, a case review was initiated. Those cases in which a levy source was identified had a Notice of Levy manually prepared by a clerk and mailed to either a bank in which the taxpayer held an account or to the taxpayer's employer. Cases where adjustments to the tax liability were necessary required either phone or personal contact with the taxpayer. Again, a number of paper forms, documenting the adjustment, were needed. Those taxpayers experiencing financial hardship, and thus unable to pay, were encouraged to visit the local IRS office to provide a financial statement and work out long-term arrangements to reduce the outstanding tax debt.

The atmosphere in which this process was accomplished was hectic, to say the least. The Contact Section handled both in-coming and out-going phone calls as well as personal interviews with taxpayers who came into the office. There was a constant bustle of activity revolving around a high degree of employee interaction.

Case material was frequently needed so that history sheets could be reviewed to assess what actions had been taken in the past, to add additional information, and to determine what disposition of the case should be made in the current situation. Frustration levels among employees ran high. Cases were frequently misfiled with a great deal of effort put into locating cases where the taxpayer called or came into the office. All case histories were handwritten (sometimes illegibly) and all documents (e.g., letters, levy notices, adjustment notices, etc.) were manually prepared. The two photographs below illustrate a typical collection of records files in such an operation (Figure 2-3).

The Collection Office Function was a highly labor-intensive operation, placing high demands on the employee. The Revenue Representatives were often rotated between

Figure 2–3. Typical scenes in the Collection Office Function.

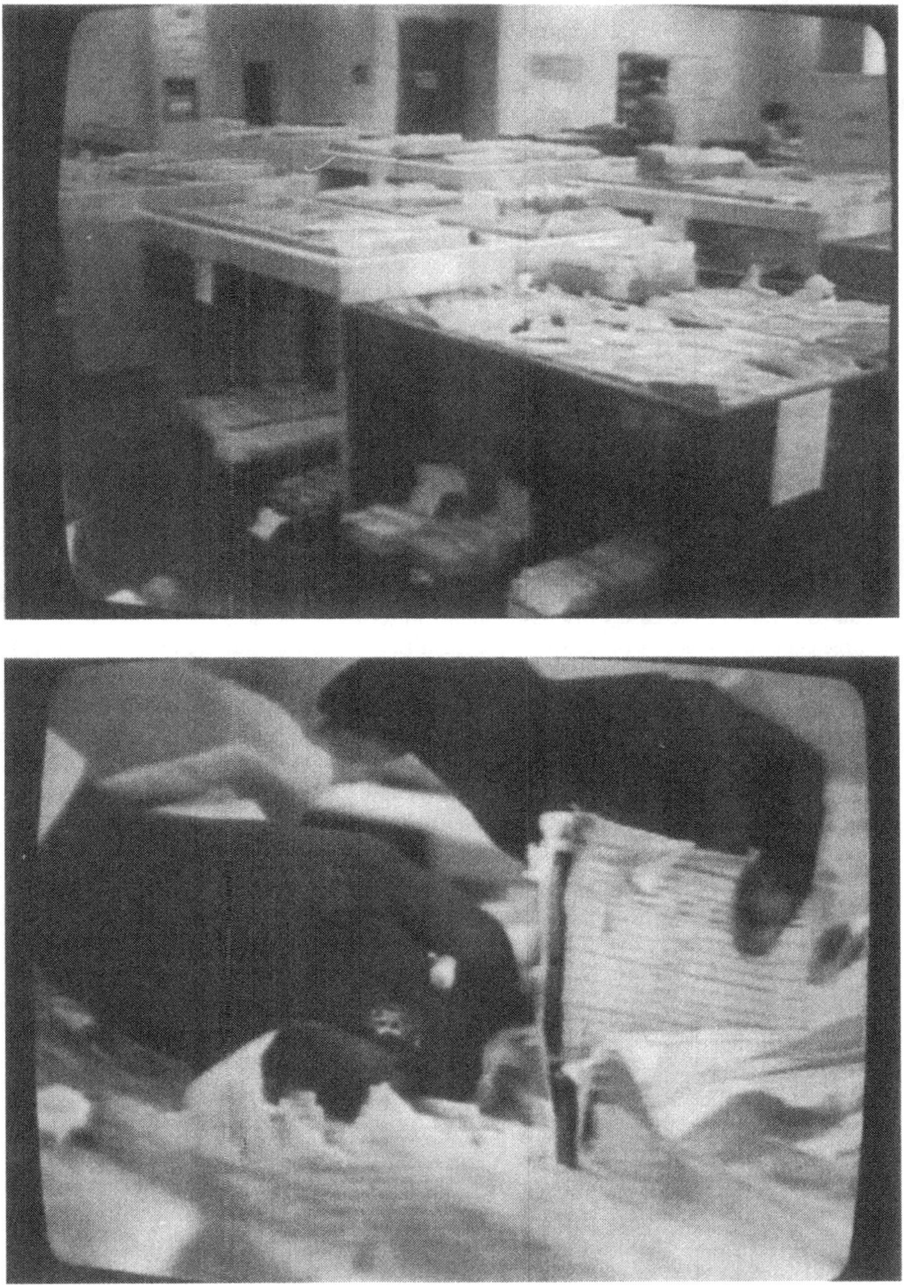

telephone contact and dealing with walk-in assistance. Frequently, if the Revenue Representative had had phone contact with a taxpayer, he or she would handle any face-to-face contact and was able to track the case from beginning to conclusion. More experienced Revenue Representatives were assigned more challenging tasks. For example, more complex business returns (as opposed to individual taxpayers) were assigned to the more advanced representatives. This policy was approved of by the employees since they were often able to follow a case from beginning to end, something that was not always possible with other cases.

The Tax Account Representatives (TARs) were responsible for the more technical reviews and work. For example, they were involved in tax adjustments, resolving accounts with the IRS master data files (called IDRS, Individual Data Retrieval Service), and handling the 'first read operations'. The first read operation was a massive paper activity requiring the TAR to review all case documents, taxpayer correspondence, Revenue Representative reports and reports from the field branch officers. Correspondence and documents received from the taxpayer required association with the case documents. Here again, a great deal of frustration was encountered when case material could not be readily located, was out of date, was partly missing or written illegibly. Quite often, inappropriate actions were taken due to such limitations.

The Process Reviewers/Clerks reviewed each batch of Taxpayer Delinquent Accounts on a 15-day cycle. Based on this review, the case was forwarded to the next appropriate action based on the history sheet of the case in question. They also had the important task of associating and compiling all paper documents that were relevant to the case. These documents included such things as responses from employers to levy requests and efforts to locate the taxpayer (called skip-tracing). When cases exceeded the 15-day period, the Process Reviewers batched them together.

Because of the complexities of the filing and retrieval systems, it was frequently difficult to prioritize the case work assignments in terms of the dollar amount owed. The three general categories of returns were identified as the Individual Master File (IMF), the Business Master File (BMF), and Overage. Within each category, cases were worked straight through. As a result, a $50000 IMF case was treated with the same priority as a $500 case. High dollar cases, however, were time and date stamped as they were received so that, hopefully, their determination could be made in a timely fashion.

The work environment was usually crowded, noisy and messy. There were no sight or noise barriers between desks and the work area was cramped to begin with, the noise levels were generally high. Those areas set aside for taxpayer walk-ins were generally filled with people waiting to be helped. Interviews with the taxpayers were frequently held in a nearby area, within sight and hearing distance of those waiting to be helped.

The turnover rate in the Collection Office Function was high, approaching 40–60% per year in some Districts. The office was, in many ways, a pressure cooker, with high demands on physical and mental stamina. As it housed the greatest number of low-grade employees in the District Office, the COF was often a fertile hiring ground for other divisions and functions. Revenue Representatives and Tax Account Representatives often moved over to become Revenue Officers (i.e., to the Field Branch). Many made outstanding contributions in that occupation and advanced to higher management positions.

Although the work was demanding and the working conditions were frequently frustrating, a good deal of fraternity existed among the employees and their supervisors. The mutual respect and group loyalty were based on shared experiences and the need to depend on each other to get the job done under less than ideal conditions. Supervisors were generally seated very close to their section. One-on-one sessions between supervisor and subordinate were the rule rather than the exception. Because of the open space office layout and the physical closeness of the supervisor to his or her employees, communication levels were high, frequent and direct. Co-workers frequently shared case information and problems, seeking advice and guidance. Newer employees were quickly brought 'up to speed' when placed in this type of familial, informal, on-the-job training. Little, if any, grade distinctions appeared to exist between technicians and clerical staff. Lunch areas, break periods, gossip, concerns and personal relationships were shared with little respect to job classification differences.

3. The impetus for automation

The Collection Office Function operated in a paper-bound environment with heavy emphasis on manual record review, storage and retrieval. In this sense, it was not unlike many other functions within the IRS, or other federal agencies for that matter. To say that it represented an inefficient use of manpower would be an understatement. To say that it was understaffed compared to the volume of transactions required would put it in a situation common to many other functions.

Under a different set of circumstances, a typical administrative action to deal with this situation might have been to reorganize the reporting relationships, increase the number of staff available, shift review or contact responsibilities from one function to another, increase the threshold level for cases requiring review and follow-up, or similar actions. This is not, in fact, what happened. The organizational reaction to the situation in the COF was the first large-scale incorporation of modern computer technology as a replacement for manual and paperwork systems.

The foundations for the movement to automate the Collection Office Function can be attributed to several factors. First, the nature of the function itself directly influenced the transformation to an automated system. In essence, the function was a debt collection agency. It brought in real cash. The fact that an overburdened and relatively inefficient operation resulted in the loss of revenue to the Federal Government was directly observable and measurable. During the period prior to the implementation of the Automated Collection System, approximately $20.5 billion were in the IRS's accounts receivable inventory. The first year of operation of the ACS was projected to cost almost $37 million followed by a second year of operation budget of less than $10 million.

The Congress of the United States, which provided the monies to automate the COF, was confronted with the argument that an investment in automation of this function would result in increased returns. The importance of this fact is clear when compared to other functions and government operations which produce less tangible 'service' functions. It is easier to justify the large investments in money and time for an organizational change when the benefits of such can be counted in cash return on investment rather than faster, better service.

Second, the case load of the Collection Office Function was increasing dramatically. Uncollected dollars were beginning to run into the billions with little hope of eventually catching up, much less getting ahead of the cases in inventory.

Adding to the increasing burden of an ever increasing inventory of cases to be processed, the efficiency of the COF was hampered by an outdated phone contact system which required manually dialled calls by the person working the case, a very time-consuming and expensive operation. The paper files, manuals, reference materials, directories, indexes and reports which were required to accomplish the collection activities were stored (or mis-stored) all over the office. Storage space itself was at a premium. The entire operation was a fertile ground for automation.

These facts must be coupled with the increasing indebtedness of the US Government resulting from years of budget imbalances. Since no tax increases were instituted under the Reagan administration, the potential for increased revenue generated by a more effective collection operation of existing debts was extremely desirable politically.

Third, the technology was already in existence to make the switch over from manual to automated systems. Proven computer hardware and telephone technology had been in use for several years in the private sector and employed for debt collection. Software, while requiring major modifications to comply with IRS information needs, was available in almost an 'off-the-shelf' fashion. In short, the technological problems, which might have been almost insurmountable if the system was to start from scratch, were more easily overcome due to the existence of a model developed by private industry.

Fourth, the need to incorporate new, automated work technologies was promoted by key individuals in positions of authority. From the Commissioner of the Internal Revenue Service to the Assistant Commissioner for the Collection Division, to the Regional Commissions in which the automated systems were first installed, the necessity to bring the IRS up to date with modern technology was advanced.

With the advantage of hindsight, it might be argued that, had any of these four major factors not been present, the Automated Collection System would not have come into being. Certainly, the need existed, the technology was available, the resources and motivation were present. All four aspects came together at the right time to produce what was to become one of the most successful federal automation efforts to date.

Chapter 3
The Politics of Automation

D. James Lantonio*

Preface

In any organizational change effort, there are a few key players who are instrumental in initiating or directing that effort. Usually, these people are those who determine the 'implementation climate' or influence the organizational values which trickle down within the organization. The IRS automation effort was, of course, no exception. Such formal and informal philosophies as, "We don't have failures in the IRS, only varying degrees of success", "The future is now", "Firm but fair", etc. were frequently translated into day-to-day procedures and practices.

From an operational point of view, the political and executive personality factors which determined such implementation values were frequently hidden from view. Such privy knowledge was usually limited to those directly involved. This chapter was written by one of those individuals who had a direct part in determining such values. His candid insights into the political aspects of such a change expose a side to the implementation that few saw or understood.

Mr Lantonio was the Assistant Commissioner for Human Resources during the period in which ACS was implemented. As such, he was responsible for managing such implementation activities as training, personnel adjustments, union relationships, etc. It was also by his direct action and sponsorship that the Human Resources Technology Staff was founded to promote human factors considerations in the introduction of the ACS.

The Internal Revenue Service made a strong commitment to its data processing activities when it introduced computers into the work environment in the early 1960s in order to process tax returns more efficiently. The equipment installed at that time was state-of-the-art. The IRS was envied by the rest of the federal community as an innovative, creative forerunner in the use of advanced technology.

By 1980, the Service was still basking in the reflected glory of the 1960s. The equipment being used had changed little from that originally purchased in the '60s, and myriad problems had begun to surface. The United States Congress had consistently refused to provide additional funding for new equipment, so 'band aids' had to be applied to those components of the system bleeding most.

One of the major reasons for continual problems was the strain on the system's capacity. Although originally designed as a transaction system in the late 1970s, much of the computer capacity was being devoted to management information systems and functional support work for the individual program areas, such as the Collection Division.

For many years, the Collection Division had been plagued with an increasing workload and reduced staffing. This trend was not limited to the Collection Division. Other functions within the IRS were also faced with this problem.

---

*Former Assistant Commissioner for Human Resources, The Internal Revenue Service

In the late 1970s, much of the initial processing of a delinquent account was done through manual office procedures. A clerk would mail the initial request for payment letters in batches of approximately 500 accounts. During the following 15 days, incoming mail would be associated with the case files, followed by a full review two weeks later. Appropriate follow-up letters would be sent if the account issues were not satisfactorily resolved. This manual process would continue until the remaining 'hard core' cases were transferred to the Field activity, where a Revenue Officer would contact the taxpayer in person.

Staffing reductions contributed to the backlog within the Collection Division. Letters were not being sent in a timely fashion; follow-up actions were delayed; and collections, in general, began to decline. The accounts receivable inventory continued to skyrocket as inflation escalated. An unhealthy attitude began to be manifest among a small, but vocal segment of the population—the late 1970s saw the emergence of the 'tax protester' movement, a movement committed to the non-payment of income taxes, even if this entailed violence.

Top management within the Collection Division recognized that automation was the only salvation for the Function. But there was little hope that Congress would appropriate funds for such a venture. The Service had requested, through the budgetary process, sufficient funds to completely revamp its data processing system. The proposed project, known as Tax System Redesign, envisaged decentralizing the Taxpayer Master File among the ten regional Service Centers. This approach would have allowed the workload to be redistributed, permitting other functions within the Service to take advantage of the computer capacity of the main transaction system without causing a degradation of tax returns processing.

Congress and the right-to-privacy advocacy groups were still sensitive in the aftermath of the Watergate scandal. There was very little enthusiasm for (and a great deal of antipathy towards) any additional access to the private lives of the population. As a result, repeated requests for additional funding fell upon deaf ears. There seemed to be no best solution, so creative managers devised second-best or halfway measures.

During the course of any given year, lags in the processing of tax returns occurred. As these lags arose, functional managers began to beg, borrow or steal the temporary excess capacity of the main computer system. Promises of 'let me run this program for only eight weeks' were grudgingly accommodated—only to have a multitude of six or eight week programs embedded in the main-frame forever. For, once a program was installed, it was rarely removed. A useful image is that of a heavily laden Christmas tree without a single branch left on which to hang an ornament, but with a line of children waiting to put just one more ball on the tree.

The result of increased workload and diminishing computer capacity was that, by 1980, the main transaction system could no longer accommodate the processing of tax returns being filed. The plethora of functional systems and the operating programs that went along with them consumed so much memory that there was every likelihood that the total system would crash.

Additional funding was secured from Congress, not to replace the existing system, but merely to add some additional equipment—to continue applying band aids. Congress believed the funding it was providing would match the growth in the number of returns filed. The IRS management knew it would only be enough to keep the system at the same level. Only as a result of the Grace Commission's in-depth analysis of the capacity issue did anyone outside the IRS become aware of the magnitude of the problem. The Grace Commission study, an independent study on the workings of the federal bureaucracy headed by W. R. Grace, made many recommendations to streamline the public sector and improve its efficiency.

This then was the data processing world of the IRS when Roscoe L. Egger, Jr was appointed Commissioner in 1981. Egger brought to the IRS a new dimension in executive leadership. Since President Truman's reorganization of the Service in 1952, management of the day-to-day operations of the IRS had been vested in the career bureaucracy. The Commissioner was the only political appointee and, in most cases, the person selected for this position was a well-known and respected attorney. The day-to-day oversight of the organization was delegated to the Deputy Commissioner. The deputy, in effect, was the Chief Operations Officer. The Commissioner usually involved himself in dealings with the Treasury Department over the implementation of federal rules and regulations.

Egger was different from his predecessors in that position. Although well-known and respected as both a tax attorney and a certified public accountant, his true interest lay in management. As a managing partner with one of the Big Eight accounting firms, Price Waterhouse, Egger had devoted the last years of his career there to developing a strategic planning process for the firm. He brought to the IRS a new vision of the future of tax administration and hoped to spend his time with the IRS implementing that vision.

After a few short months in office, Egger had drafted an agenda for his tenure. He recognized and appreciated the sound management principles the Service was operating under. He took comfort in the fact that the IRS was recognized as the best managed federal civilian agency. His aim was to ensure that legacy into the 21st century.

Egger focused his attention on three principal areas, all interrelated. His overall thrust was: (1) to establish a continuing strategic planning process; (2) to couple the strategic planning process with a complete revamping of the data processing functions within the Service; and (3) to restructure the organization to allow much more decentralization of power. Egger was convinced that managers must be given the delegated authority to make decisions at the lowest possible level in the organization. Coupled with this decentralization of authority was the belief that managers should be held accountable for their actions. His initial impression, which proved valid over the course of time, was that no real decision-making occurred in the field organizations. Field managers were, in effect, paid to monitor, not to manage. Egger tried to attack all three issues simultaneously.

It would be unfair to say that the organization did not attempt to co-operate. The Service has always prided itself on its willingness to support the wishes of the person appointed to head it in so far as those wishes did not violate the law or compromise the public's confidence in the integrity of tax administration. Unfortunately, what Egger articulated and what the organization heard were two different things. Egger was preaching a new vision of management, the Service was hearing a minor tinkering with the current way of doing business. When Egger spoke of a strategic planning process, the organization heard modifications to the long-range planning process. When he spoke of a plan that would drive the budget planning system, they only heard a multi-year budget planning approach. When he talked about an Information Resource Management function, they heard further centralization of the tax processing system. And finally, when he spoke of decentralized decision-making, they heard tighter controls at the Regional level. Needless to say, the first 18 months of Egger's tenure frustrated both him and his career managers.

The reaction of the Washington bureaucrats was understandable. Many had spent more than 30 years in an organization that was steeped in tradition and in a culture that rivalled the Federal Bureau of Investigation's in self-esteem. The organization had performed very well; it had stood up strongly against attempted inroads by political ne'er-do-wells; the public maintained a relatively high degree of confidence in the way the tax system was being administered. With this foundation still germane, why change? Fortunately for Egger, there were some believers among his field executives. One, in particular, heard the word and was willing to march to the beat of a different drummer.

The Regional Commissioner in Chicago had long been concerned with what he perceived to be the dwindling respect the average American showed for the voluntary payment of taxes. Careful study of compliance levels revealed a perceptible downward trend in such compliance. The reduction in the number of tax returns being examined because of reduced staffing levels; the lengthening of time between notices being mailed on delinquent accounts; the reduced number of special agents available to work on general tax fraud as opposed to 'sexy issues', such as narcotics and money laundering—all persuaded the Regional Commissioner that Egger's view that the old ways of doing business had to be abandoned was right. As the Regional Commissioner pondered the dilemmas facing the Service, he decided to test the waters and see whether Egger would indeed support a new way to do business. He focused his attention on the Collection Office Function.

As already mentioned, the deluge of paper from the increased number of accounts receivable and the loss of irreplaceable employees had created a problem of national concern. An effort had been made during the late 1960s to automate part of the function as an add-on to the major transactional system. This adjunct was the IDRS (the Integrated Data Retrieval System). IDRS had significantly improved the Service's ability to resolve disputed collection cases. By 1980, however, IDRS was stretched beyond its capacity, especially since the mainline tax processing system on which it was based was about to be submerged under its own weight.

The Midwest Regional Commissioner in Chicago decided to look for a system, similar to IDRS, but with its own, separate data base, one that would feed the mainline system instead of being fed by it, as was the case with the IDRS. In Minnesota, he found a small collection service bureau that was using a computerized system to augment its workforce in collecting delinquent accounts for its clients. As the Regional Commissioner and his staff studied the service bureau's procedures, they became convinced that a comparable system would not only be feasible within the government structure, but would be the salvation of the IRS's collection program.

The Regional Commissioner and his staff quietly spent months studying the alternatives and designing a proposal to be submitted to the Director of the Collection Division, the Assistant Commissioner for Compliance Matters, the Deputy Commissioner, and the Commissioner. This group of individuals formed the power structure in Washington whom they perceived as able to say yea or nay to pursuing such a system further.

The proposal was presented at a meeting in Washington. The Regional Commissioner and his staff made a two-hour presentation, supported by a cost–benefit analysis, demonstrations of how the system was working in the private sector, and projected yield the Service would reap over a five and ten year basis. These factors provided the sound, logical reasons why it was important to pursue the idea. Although the skeptics in the room were impressed, myriad reasons why it would not work were given.

Egger, becoming more and more frustrated, finally ended the meeting by stating that the idea was one of the more creative and innovative he had heard since joining the Service. He directed the Deputy Commissioner to put together a group to begin to develop the presentation for submission the Treasury Department and the Office of

Management and Budget.

Within weeks, the political infrastructure of the organization began to assert itself. Executives staked claims to the project. Each found a legitimate reason to be involved, and each thought he should control the project. The Assistant Commissioner for Computer Services claimed ownership since the system under consideration was a computer system. The Assistant Commissioner for Returns and Information Processing claimed ownership on the basis that tax processing information would be put in the system. The Assistant Commissioner for Support Services felt he should be in control of the project, since the system would involve massive procurement, which fell under his jurisdiction. The Assistant Commissioner for Examination, although not seeking the chairmanship, wanted to be actively involved since he was independently exploring a similar system for examining tax returns. The Assistant Commissioner for Collections unequivocally stated that it was his system. The Regional Commissioner for the Midwest, on the other hand, made it quite clear that it was his idea and that no one in Washington was going to wrest control of the project, only to have it die like so many other good ideas that had arisen in the past.

Ultimately, a compromise was struck. The proposal was officially designated as an automation project. A project manager from the Data Processing Division was appointed, backed up by an assistant project manager from the COF. Overall supervision of the project was vested in the Automation Implementation Control Group, which would report to the Service's overall Automation Policy Review Board. The Automation Policy Review Board was composed of three Associate Commissioners, two Regional Commissioners and the Deputy Commissioner as Chairman. The Commissioner was an *ex officio* member. The Automation Implementation Control Board, on the other hand, consisted of all executives who could even remotely be involved in the project. The membership was as follows:

Associate Commissioner, Data Processing (Chairman)
Associate Commissioner, Compliance
Associate Commissioner, Policy and Management
Assistant Commissioner, Computer Services
Assistant Commissioner, Planning, Finance, and Research
Assistant Commissioner, Examination
Assistant Commissioner, Collection
Assistant Commissioner, Support Services
Assistant Commissioner, Human Resources
Regional Commissioner, Midwest

Needless to say, the political maneuvering during the early stages far exceeded the actual amount of work accomplished. As time progressed, however, other demands on the executives forced them to allow the project manager and his assistant some discretion to develop a comprehensive plan for an RFP, Request for Proposal.

The concept of this project sent shock-waves throughout the IRS organization; it was unlike anything the Service had ever undertaken. There was even talk of going to a well-known vendor rather than the 'lowest bidder' to manage the design and implementation of the systems. There was a sense of urgency unparalleled in the history of the Service. Early on, it became apparent that the only way the proposal would ever survive the Executive Branch review (Office of the President) was if a demonstrable payback could be shown while Reagan was still President. The administration's desire to balance the federal budget became a window of opportunity through which to sell the concept of an automated collection system.

Creativity was at its peak during the next 6–9 months. Analysts spent every working hour looking for loopholes in the Government Supply Agency's (GSA) cumbersome procurement policies and regulations. The 21-month procurement cycle was reduced to the bare minimum. Marketing strategies were reworked and refined to ensure wholehearted endorsement by every level as the proposal made its way through the various levels of the Executive Branch of government.

Once it became apparent that there was a chance the proposal would become a reality, the rest of the IRS organization began to maneuver for control. All the functions looked for ways to tap into the new system. The prospect of the heavily laden Christmas tree loomed all over again. Fortunately, the Automation Policy Review Board mandated that there would be no tinkering with the original design and that the system would be used by only a single function, an automated collection system.

In due course, the procurement process ended, and a contract was awarded to Gulf Coast Services (GC) of Houston, Texas. The GC Services bid was based on a joint venture combining GC Services' collection expertise, IBM hardware, Arthur Anderson Corporation's software development and Rockwell International's telecommunications capabilities. For the first time in decades, the IRS was to have a state-of-the-art computer system to support a single functional activity. The only fly in the ointment was that both the President and the Congress expected a quick payback for appropriations invested in the system.

Plans were immediately developed to have pilot sites on-line within 4 months from contract award. The overall, very ambitious program called for 20 operating sites to be in operation in less than 18 months. The briefness of the time span for implementation was the only way the payback could be achieved during the tenure of the Administration.

As the planning process advanced, some members of the Automation Implementation Control Group realized that a lot more thinking and strategizing had to be devoted to the 'people' issues. In effect, there were two basic issues that had to be addressed: the implementation of the new system could conceivably make several thousand employees redundant; and there would be a need for an entirely new way of training and orienting the remaining employees to work in an automated environment.

Although the Service had over 20 years of experience in managing a highly automated workforce in its Service Centers, the ACS system would be completely different. In the Service Center operation, the workforce was primarily part-time employees, who worked only during the peak of the tax season. The jobs of these Service Center employees were very restricted; they were to enter, directly into the computer database, information from standardized tax forms. This did not challenge their abilities. The Automated Collection System would do just the opposite.

The first issue, employee redundancy, was a particularly sensitive one. Most rank-and-file IRS employees are represented by the National Treasury Employee's Union (NTEU), an organization known for its militancy. The relationship between NTEU and IRS management had been marked by years of bitterness. It had only been within the last three or four years that some improvement in the relationship had occurred. Negotiations over the treatment of the displaced employees would provide a true test of the new philosophy of management–labor collaboration to solve problems.

Both sides worked exceedingly hard for three months to hammer out an agreement that was both equitable to the employees on the issues of redevelopment and training and also afforded management the flexibility it needed to implement the new program.

The second issue, the reorientation of employees to the automated system, proved to be the thornier problem. It was here that the management found itself pitted against itself. Early on, the Human Resources people raised questions about the alienation factor that most other large organizations had faced when such large-scale automation projects had been implemented. The Human Resources organization began to develop long-term programs to deal with the myriad issues requiring attention. The ACS project office and the Collection organization, on the other hand, resisted any attempt at what they considered to be interference. Their initial defence was to stonewall any activity that might jeopardize meeting the program completion dates. They were little concerned that independent studies showed the initial turnover rates for similar ventures to be more than 200%. Nor were they overly concerned that there was an insufficient base of employees with the skills needed to work in this new environment, nor that a long lead time was required for really adequate training. The overriding IRS objective was to get the system up and running by a specified date. They would brook no interference, especially from the Human Resources part of the organization. It was, in their view, meddling, another attempt by someone else to influence their system.

The Human Resources organization, on the other hand, had a genuine concern that if the people issues of the new system were not handled properly from the outset, it would be doomed to failure. The recommendation was made to invest the extra weeks or months on the front end, to ensure success over the life of the system, rather than to risk the system's failure.

The Automation Implementation Control Group, although sympathetic to the concerns of the Human Resources organization, was loathe to allow anything to interfere with adhering to a very rigid time schedule. Ultimately, the Board grudgingly agreed that the Human Resources people must collect some data at the pilot sites, but only so long as the data gathering did not interfere with the overall program implementation plan. This compromise was the genesis of what became known as the Human Resources Technology Group (HRT), a small group of experts in industrial psychology, organizational research, training and job design. In order to allay the fears of the organization, the group was initially kept *ad hoc* in nature. It had no form or authority structure in the traditional IRS meaning of the word. It had no formal leader, and its mandate was free-flowing. The only reporting relationship for the group was directly to the Assistant Commissioner for Human Resources. Meetings were held with the Assistant Commissioner either individually or as a group, depending on the particular issue they were working on.

The HRT Group met resistance at every turn save in the field front line. The National Office functional representatives demanded to review and approve any action proposed by the HRT staff to ensure that it would not impede progress towards the ultimate implementation objectives. If the HRT staff proposed the administration of a questionnaire, it could only be of a certain length and duration so as not to interfere with 'the real work'. The same bias applied to training employees and managers. To the functionalist, the issue of training was pure and simple. Only one week before going on line, all employees should receive training on the technical aspects of the system. They would be placed in a classroom environment where they would be lectured to concerning system operations. They would not get any 'hands-on' experience with the system until the week following the training, at which time they were to report to the new automated office site. Regional officials reacted to this approach similarly. They found themselves pressured to deliver an implemented system to the National Office within the very rigid time frames established. The last thing they wanted was to be perceived as a bottleneck in implementing the system.

It was only at the District level that a different attitude emerged. Here the theoretical work system had to come to life. Here the people were real and not simply concepts or numbers. It was at this level in the organization that the members of the Human Resources Technology Group found a welcome. The group's work became almost covert. The limited involvement by the other Human Resources components (i.e., personnel, training and development) at the top of the organization continued, while there was a flurry of activity directly between the District Offices and the HRT Group. True consultation began to take place. Eventually, the District Offices exerted enough pressure to allow greater and more substantive involvement in the implementation by the HRT group.

Three years have passed since the Automated Collection System first went online. During that time, a similar system has been designed for the Examination Division of the IRS. A complete redesign of the tax processing system is underway and a number of smaller systems are being implemented. In 1986, Roscoe Egger retired as the Commissioner of the IRS. The world of the IRS today looks nothing like it did in 1980. Through slowly increased awareness, uneasiness and outright hostility towards the role people play in organizational growth and development has given way to a fuller understanding of the need to involve people at all levels in the process.

The attitude of executives has changed significantly, but the process has been painfully slow. In order to effect such a change, there have to be more leaders who are willing to take an often lonely risk and rally others to join them. A Senior Executive once remarked that trying to change an attitude within the IRS is like changing the direction of the Queen Elizabeth 2 in the Potomac River. The Queen Elizabeth 2 is indeed changing direction, slow as it may be.

Chapter 4
The Human Resources Technology Staff

Introduction

The Human Resources Technology Staff is probably unique among federal departments. Its existence and effectiveness can most likely be attributed to the Office of the Assistant Commissioner of Human Resources. What started out as a temporary conference assignment for two mid-level staff analysts became, over the course of three years, a full-time permanent staff of ten professionals and support staff whose function was to assure the inclusion of human factors considerations in all IRS technology related projects.

1. History and functions

The history, structure and functions of the Human Resources Technology Staff (HRT) can be traced back to the 'Joint Conference for Executives' in the Fall of 1982. The purpose of this Joint Conference is to draw together all senior IRS executives for an annual review of major programs, issues and future directions that the IRS is taking. Presentations are made in each of the major functional areas by responsible officials.

The then newly appointed Assistant Commissioner for Human Resources, D. James Lantonio, was convinced that the trend in human resources was going to center around the implementation of computer-mediated work technologies. His concern was that little planning or thought had been given to the consequences of these new technologies on the impacted personnel. Outside the traditional training requirements and such personnel activities as writing new position descriptions, recruiting new employees, and union negotiations, little in the way of human factors, new management styles, stress management and organizational change strategies was being considered. The theme he proposed for the Joint Conference involved just such an emphasis.

To accomplish this two staff members, one an industrial psychologist and the other a management training expert, were temporarily assigned to establish a program to introduce executives to the concepts and consequences of human–computer interaction. Three nationally known experts in socio-technical systems design were invited to give presentations during the conference on this topic. For the most part, these presentations were well received and served to heighten the awareness of a traditional body of executives to the needs of employees and managers as the IRS moved to automate many of its functions.

The major focus of automation for the IRS at that time was the forthcoming Automated Collection System implementation. To build on the issues raised during the Joint Conference, the Assistant Commissioner established a temporary Task Force composed of four members, the two already identified plus a future ACS Branch Chief from the Indianapolis Call Site and a Section Chief from the Position Management Branch of the Personnel Division. Identified as the Human Resources Technology Group, these individuals were given the task of coming up with a practical plan to measure and promote human factors considerations as part of the ACS pilot test in four call sites that were to go on-line within five months of the group recommendations. Three weeks were allotted to prepare a series of proposals for presentation at an ACS conference to be held in Chicago the following January.

The initial series of activities, as presented to the Assistant Commissioner for Human Resources, formed the basis upon which the group was to develop its philosophy and activities. This twelve page report is presented as Appendix 1.

In summary format, the objectives of the HRT for the fiscal year 1983 were as follows.

Objective I. Data collection

By the end of the Automated Collection System test period in the pilot sites of Chicago, St Louis, Indianapolis and Nashville, human factors data will be collected by the following methods:

1. Employee questionnaire.
2. Exit interviews.
3. Employee turnover analysis.
4. An on-line employee feedback system.

Objective II. Data analysis

By the end of the pilot test period, data collected will be analyzed to assess the following issues:

1. Issues arising out of the use of the automated equipment.
2. Adequacy of training with respect to the changes resulting from automation.
3. The degree to which ergonomic needs were being met in the pilot sites.
4. The need for use of innovative organizational effectiveness approaches.

Objective III. Data distribution

Data will be periodically furnished to the Human Resources management in the National Office, National Office Collection Division, ACS National Office project managers, pilot site regional officials, National Treasury Employee's Union officials and others, as appropriate. The HRT will serve as an information resource throughout the implementation of the ACS project.

Objective IV. Orientation

1. Provide guidelines and information on suggested employee orientation activities.
2. Distribution selected readings and computer literacy information to call sites for redistribution and employee reference.
3. Prepare suggested remarks for a 'kick-off' orientation module to begin training.

Objective V. Training

Training will be developed for ACS employees (supervisory and non-supervisory) in the following areas:

1. Managing stress and managing change.
2. Problem finding and problem solving.
3. Job redesign awareness (managers only).
4. How to use the on-line electronic feedback and questionnaire developed by the HRT group.

The primary focus of the HRT activities with respect to the implementation of the ACS thus fell within three major areas: (1) collecting information regarding the impact of automation on personnel; (2) acting as a resource to the ACS to provide guidelines and information on the human factors of automation; and (3) designing a series of training seminars focusing on preparing employees and managers to deal with the new automated work environment.

Such a series of activities required additional staff. A facilities manager, an ergonomist, a doctoral student in organizational communication, a person versed in public administration, and an individual experienced in all aspects of the Collection Division were, at various times, added to the staff of the HRT.

The HRT Group grew in size and activities. By March 1984 it had become a permanent staff reporting directly to the Assistant Commissioner of Human Resources. With respect to the technical areas of specialization, the HRT structure was both multi-disciplinary and cross-functional. This is illustrated in Figure 4–1.

Figure 4–1. The multidisciplinary and cross-functional areas of specialization in the Human Resources Technology Staff.

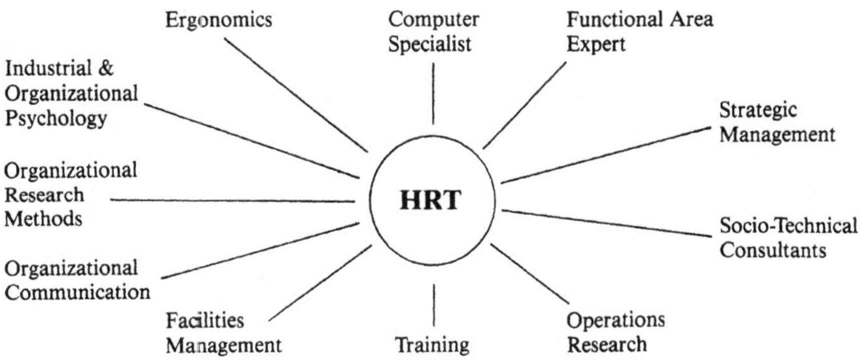

The mission statement of the HRT Staff provided the framework within which it would operate:

> Provides advice and assistance to Service management in planning for and dealing with the human factors associated with the introduction of automated technology into the work environment. Provides consultation services to executives and organizations on the application of interpersonal communication, management theories and practices to specific human factors situations such as consideration of ergonomic factors, providing for end user participation in system planning and design to ensure that human factors needs have been addressed, determining the impact of technology on organizational structures, policies, procedures, job and career ladders, and managing the process of change and job related stress. Develops and conducts human factors research to support consultation activities, and to increase the Service's knowledge of the state of the art. Serves as a principal point of

contact for all IRS organizations desiring information on human factors activities and studies within the Service, in other Federal agencies, private sector organizations, and academia, and serves as a repository of information on ergonomic actiivities. Disseminates information as requested or as the need is perceived. Coordinates and collaborates with other functional organizations as appropriate, including Personnel, Facilities Management, and Training and Development Divisions, and appropriate offices of the Assistant Commissioner (Computer Services) on matters dealing with the human factors associated with implementing automated systems. Maintains liasion with other government agencies which are involved in similar activities.

In addition to its unique focus on the human factors of automation, the HRT Staff was unique to the IRS in three other respects. First, for the majority of its existence it operated as a leaderless group. No individual was assigned a supervisory status. The charge of the group, as defined by the Assistant Commissioner, was to develop a structure which best suited its people and functions. A collegial model evolved as a natural consequence of the personalities involved.

This type of organizational structure proved to be confusing to many non-HRT individuals. Since each member of the HRT Staff had the authority to speak for the rest of the group and make commitments in the name of the others, more traditional IRS managers sometimes had difficulty in dealing with the rather nebulous authority structure of the HRT Staff. To overcome this, one member was appointed as a 'coordinator'. This role was not to act as a leader and decision-maker but rather as a spokesman for the HRT Staff with the rest of the IRS and to keep track of the many administrative details that a traditional supervisor would be expected to accomplish. In no way did this individual exercise more or less authority within the group than any other member. Important issues or decisions were brought up for group discussion and vote prior to any action.

Another way in which the HRT Staff differed from the more traditional IRS Branch was in the nature of some of its activities and reporting relationships. HRT Staff members, in accomplishing their data collection and consultant role to various ACS call sites were not obliged to issue formal reports of activities and accomplishments. In practice, this meant that any relationship established with a specific District Director or ACS Branch Chief was considered non-public. A study conducted for a given call site, for example, was considered privy information between the HRT Staff and the call site official. Any reports or recommendations were issued only to the requesting official. Only upon his or her approval could this specific information be disseminated. General issues and conclusions, without reference to the source of the information, were frequently communicated to other sites. Although resulting in much aggravation from other sources, this policy provided a great deal of goodwill among the sponsoring officials.

The reports made to the sponsoring office, either in the form of a verbal briefing or written summary, were often the basis upon which actions were taken. Generally speaking, the more concrete, immediate, and practical the issues raised, the greater the likelihood they would be addressed at the local level. Those problems involving more abstract, long-term or wide-spread concerns were much less likely to be addressed. It was usually easier to apply a band-aid rather than major surgery.

The third significant manner in which the HRT Staff differed from the traditional organizational structure related to the cross-functional authority with which it operated. Reporting directly to the Assistant Commissioner provided the HRT Staff with the assumed authority to interact with all levels of Field and National Office executives. Little regard was paid to the chain of command or functional boundaries. This freedom to avoid much of the paperwork, hierarchical reviews and approval authorities did much to increase the responsiveness of the HRT Staff to the needs of the ACS. The ability to cut through the 'red tape' that other branches encountered allowed a turn-around time uncommon in the Service.

2. *Measuring the impact of automation*

A time frame of three months existed from the presentation of recommendations to the start-up of the first ACS call site. As a consequence of this, the initial role of the HRT Group became one of measuring human factors rather than active participation in their promotion.

Four ACS call sites were to act as a pilot test for the remaining sites. Two different vendors were selected to design and manage the design and installation of all equipment, software programs, training, and telephone systems. In essence, these two vendors were in competition with each other for the contract to implement the remaining call sites. Vendor A was to implement the Chicago and St Louis call sites with Vendor B charged with the Nashville and Indianapolis sites. All four sites were due to become operational at the same time. Due to the nature of the competition, the HRT was not to engage in any activity which might be construed as unfairly affecting one contractor in favour of the other. Additionally, the large number of IRS visitors, foreign representatives, media and contractor people on site during the pilot phase tended to create a highly charged atmosphere. It was felt that the addition of HRT people would only add to the burden of such a high profile operation.

The data collection effort planned by the HRT was thus limited to the minimal level of on-site presence. Individual or group interviews and on-site observational methods for data collection, were thus ruled out. As the most appropriate methodology, the HRT staff developed a series of questionnaires. Three types of such instruments were initially developed.

1. A self-administered exit questionnaire for employees who were in the process of leaving the ACS.
2. An electronic feedback form. This was a postcard-sized form, copies of which were to be kept at the employee's workstation. When an event occurred that had human factors implications, they were to complete the form giving a narrative description of the event. The forms were to be collected on a weekly basis and copies sent directly to the HRT Staff in Washington by means of a microcomputer located at each site. These microcomputers were specifically purchased for this electronic network.
3. A comprehensive, self-administered questionnaire designed to be answered three times. The first administration was prior to joining the ACS, the second was following a one month experience in the new system, and the final administration occurred after six months.

Of the three data collection methods, only the paper and pencil questionnaire proved to be a viable source of information. Neither the electronic feedback form nor the exit survey were administered with enough frequency to permit any statistical conclusions. The reliance on site staff to promote their use was, in retrospect, an unavoidable error. Since their primary duty was to get the call site up and running rather than admonish people to complete exit forms or fill out daily 'complaint' forms, little attention was paid to these activities. The "What's Your Opinion?" survey, however, did prove to be a valuable source of information. A copy of the final form of this survey is included in Appendix 2.

In filling out the questionnaire, all employees responses were anonymous. The questionnaire took about 15 minutes to fill out after a brief introduction concerning the purpose, union approval, and directions for completing the questionnaire was presented by a member of the HRT Staff. The questionnaire itself was composed of five main sections:

Section 1 concerned employee socio-demographic information, ACS position, prior work history and educational level.

Section 2 covered employee satisfaction with the work process.

Section 3 dealt with issues concerning the work environment.

Section 4 contained questions covering satisfaction levels with the work equipment and software.

Section 5 measured the degree of satisfaction with the IRS, the ACS, co-workers, peers, training and several other broad issues.

A quasi-experimental research design was proposed to measure the changes, if any, that could be attributed to the implementation of automation. The original data collection plan was to administer the What's Your Opinion? survey at all four pilot call sites three times, the first administration to occur in March 1983 (i.e., just prior to the beginning of ACS training), the second in June (six weeks after start-up) and the third administration in October 1983 (representing six months operational experience). Unfortunately, one of the two pilot site vendors was unable to participate due to a technical limitation on the proposed phone system and withdrew prior to site preparation. As a result of this default, the pilot test sites at Nashville and Indianapolis were not able to go on-line as planned. These two sites were brought on-line in October 1983. All four pilot call sites had been administered in the survey during March. The two late starting call sites, Nashville and Indianapolis, were not given the second administration of the questionnaire but did participate in the third administration after they had been in operation for six months. Thus, while the calendar times of administration for all four call sites do not coincide, the length of experience between administrations are very similar.

Since the four call sites from which data were collected were considered pilots for the remaining sites, the need for a comparative basis was obvious. Two Collection Office Function sites were chosen to provide a database of satisfaction levels that existed under the older, manual systems. It was recognized that many experimental design problems existed with this design but these were beyond the control of the researchers. The two comparison sites, Boston and Denver, were chosen for their similarity with the pilot sites (population served, size of operation, number of employees, etc.) and the fact that they were scheduled to become ACS sites during the later phases of the total implementation. It was hoped that neither the potential for Hawthorne effects (i.e., the potential effects of being placed in an experimental group confounding the results) in the pilot sites nor the potentially depressing effect of being phased out in the COF sites provided a significant bias to the data collected. As often happens with field studies, the choice was to collect potentially flawed data or none at all. We chose the former option with full knowledge of its limitations. The results of this data collection are presented in Chapter 6.

Data collection by questionnaire was not the only method on which conclusions were based. During the pilot period, the HRT Group met several times with implementation managers and attended staff meetings at which the progress of the implementation was discussed. Progress reports, produced regularly by the call site project management staff, also proved to be a valuable source of information regarding human factors issues. After the pilot test period, several on-site visits were made to most of the call sites. Particularly useful information, much of it forming the basis for later chapters, was collected by means of structured group interviews of employees and managers.

Taken together, the several sources of information directly or indirectly available to the HRT Staff provided a fairly comprehensive picture of the relationship between the implementation of automation and human factors. During the progress of the study, it became clear that there existed two sets of factors operating during the implementation. The first set of factors could be identified as general issues that appeared in almost all call sites. The second set of factors appeared to be those specifically attributable to the individual call site, its management, environment, history, location, etc. For the purpose here, the emphasis will be on the general issues and their implications for human factors.

Section II

AUTOMATION IMPLEMENTATION

Chapter 5
The Automated Collection System

Introduction

The Automated Collection System (ACS) was the Service's answer to the problems inherent in the Collection Office Function. In its simplest form, the ACS was an effort to convert the manual operations of information storage, retrieval, telephone contact, forms preparation and control into a computer-based automation operation.

To accomplish this, the Internal Revenue Service issued a 'Request for Proposal'. This was an invitation for private sector organizations to assist the IRS in designing and implementing the automation effort envisaged by IRS officials. More specifically, the government wanted contractors to help design an automated telephone system, the software and hardware for a database of taxpayer information, and the site requirements for such a system. Initially, two contractors made acceptable proposals to accomplish this but one defaulted due to technical problems in the automated telephone system.

The entire operation of the ACS was predicated upon the successful implementation of the pilot test at the four selected call sites. The call site pretest was designed to debug all systems prior to full implementation of the ACS which was to have been phased in over a period of one year following the pretest. In all, 22 call sites were planned to be in operation by July 1984. A period of a little over two years, from June 1981 to August 1983, elapsed from the issuance of the request for proposal until the first two pilot call sites went on line. This was an incredibly short period of time considering the complexity of the operation, the number of people involved and the myriad federal procurement regulations that had to be dealt with.

During the installation process, the overall direction of the implementation was under the control of the Automation Implementation Control Group, comprised of senior executives heading up National Office Divisions and Regional Commissioners. Following the period of implementation, the operation of the call sites was turned over to the Regional Collection Divisions for day-to-day management.

The overall success of the Automated Collection System cannot be denied. It paid for itself in cost savings and increased revenue within a very brief period after start-up. The problems that were encountered were many. No automation effort of this magnitude can be expected to be implemented without some problems. In hindsight, these problems were frequently avoidable and predictable, but given the time-frame and the rather revolutionary changes that ensued, the system is a successful one. The problems and issues that arose are detailed later in this book.

2. *The technology of automation*

From the viewpoint of the technology involved, a radical change in the physical operation of the Office Branch of the Collection Division was accomplished. Where possible, electronic operations were substituted for what were, previously, manual or mechanical operations. The two major areas in which these substitutions were accomplished were in the telephone dialing system and in the file maintenance areas.

The automated telephone dialing system in the ACS was a considerable improvement from that used in the Collection Office Function. The operator was no longer required to manually dial a number. The push of a 'next call' key on a terminal keyboard prompted the computer to automatically dial the telephone number of the next case with the highest priority. The computer program selected the calling method and telephone lines which involved the lowest cost to the IRS. The operator was no longer required to cradle the phone in the crook of the neck to have the freedom to use his or her hand. Telephone headsets were now employed which positioned a hearing and speaking mouthpiece over the operator's head.

If calls were not successfully placed, i.e., the line was busy or there was no answer, the computer program automatically rescheduled the call for a later time. This, of course, freed the operator from having to make a file notation and recall later that the call was not successful. The two photographs below (Figure 5–1) illustrate some typical operations within the new automated environment.

2. *The organization of the ACS*

Unlike the Collection Office Function, the Automated Collection System was not located in each District. In all, 22 call sites were planned. Due to the increased efficiency of operation, these 22 sites would be able to handle the workload and operations of all the previous sites.

Each call site planned for the ACS was physically constructed or revamped. In several cases, the ACS facilities were built from scratch on new construction sites. In other cases, existing IRS facilities were totally redesigned to incorporate new ACS equipment and operations. In all cases, the office furniture, workstations, terminals, computers, cabinets, break areas, security systems and work environments were specifically designed or purchased with the functions of ACS in mind.

The typical call site averaged 100 employees. In all, about 2500 employees work within the ACS. This represented a sharp reduction in the number of employees within the Collection Office Function. The issues involved in redeploying staff and case transfer will be discussed later.

In an environmental sense, the new ACS call sites differed dramatically from the old Collection Office Function. The dramatic physical and environmental differences were planned. Conceptually, the planning staff wanted to emphasize the difference between the old and new way of working.

The most striking differences were the physical changes. Since the call sites were constructed specifically for the ACS, efforts were made to establish a subdued, professional atmosphere. Air conditioning required that all windows be closed, thus reducing the intrusion of outside noise. Overhead lighting was used to eliminate dark areas and create uniform brightness. Further reduction in noise levels resulted from the use of carpeting throughout the work area.

Replacing the traditional desks were a series of cubicle workstations. Each workstation included an ACS terminal, keyboard and personal storage cabinet which contained a file drawer and small utility tray. Each employee was given a personal telephone headset which they kept with their personal effects.

Figure 5-1. Typical operations in the new ACS environment.

Apart from such small personal items as a purse, employee personal effects were not allowed at the workstation. Each employee was assigned a personal effects locker within the call site in which they were told to hang their coats, place their packages, and other personal belongings. Since no individual was to be assigned to a specific workstation, no such individual items as family pictures or memorabilia could be located at the workstation.

The desk and chair within the workstation were adjustable to suit the individual height and comfort levels of the ACS personnnel. The terminal screen was swivel-mounted and adjustable for two positions of tilt. Similarly, the keyboard was movable since it was attached to the terminal base unit by a coiled cable. In general, each workstation was surrounded on two sides by a five foot (1.60m) high partition which acted as a noise and sight barrier. The general configuration of the workstations was that of a pinwheel, although each site was free to set up the configuration that most suited its office space requirements.

Managers generally were located physically near their employee workstations but were allowed to maintain a more traditional office setting incorporating a work desk, table and chairs. These spaces were usually enclosed by movable space dividers on four sides to increase the sense of privacy while still being consistent with the open architecture of the rest of the office.

Each ACS call site was constructed with a break area. This area typically contained a refrigerator and a number of tables and chairs for lunch. Often, a microwave oven and vending machines were available to employees. Smoking and drinking were allowed within the break area. Although almost all sites allowed smoking at the workstation, few allowed beverage consumption there. Those that did required employees to purchase non-spillable coffee cups.

The work environment and emotional tone of the ACS site differed dramatically from the COF. Gone was the loud, noisy atmosphere. This was replaced by a quiet, almost subdued environment. Since personal contact with taxpayers was no longer associated with the Function, individuals and families waiting to talk with Collection Office representatives were no longer in evidence. Gone also were the huge bins of case records and filing cabinets which dominated the Collection Office Function. Since incoming telephone calls were physically separated from outgoing calls, the constant ringing of telephones was nonexistent. Electronic printers used in the call sites were enclosed by a sound barrier to reduce their effect in the work area. Since all equipment had been newly purchased for the ACS, the overall appearance of the work environment was bright, new and uncluttered. In general, many attempts were made to promote a sense of business and professionalism in the new environment without the sterility of many govenment office locations.

In a behavioral sense, communication among employees was restricted. The workstation barriers made interpersonal communication difficult. Since employees, at least in the Contact Function, were required to wear telephone headsets, the degree of physical mobility was severely restricted. Questions and discussions among co-workers, common in the Collection Office Function, were not in evidence among ACS employees. The focus, in many ways, changed from a peer-oriented work environment to that of a manager-focused one.

Another striking difference between the Collection Office Function and the Automated Collection System was the type of employment categories. During the implementation of the new systems, the National Office recommended that a staff ratio of 60% full-time permanent staff to 40% 'other than full-time permanent' staff be established. This 'other than full-time permanent' group was composed primarily of temporary full-time and permanent part-time employees. It was hoped that the temporary staff could be used on an as-needed basis to cover periods of high demand and reduced in periods of low demand.

3. *The management structure of the ACS*

The managerial structure of the ACS was a modification of that found in the COF. Since the goal of the Function was the same, the functional operations were, in many ways, similar. The stated goals of the Automated Collection System are threefold: to increase dollars collected; to increase employee productivity; and to increase taxpayer compliance.

The management structure, reporting relationships, and position titles of a typical ACS call site are illustrated in the diagram below, Figure 5-2.

Figure 5-2. An ACS pilot call site organization chart.

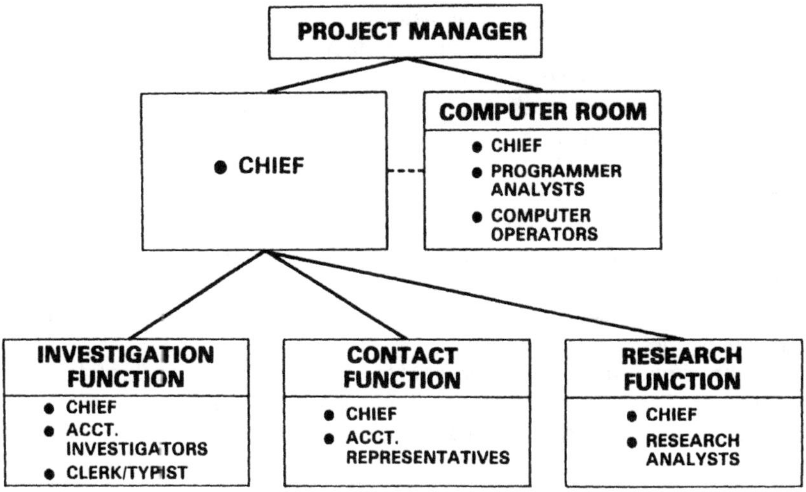

Compared to the Collection Office Function, the Automated Collection System was designed around three major sections: the Contact Section; the Investigation Section; and the Research Section. The purpose of the Contact Section was to place outgoing phone calls to taxpayers to secure delinquent returns or payment of delinquent accounts. The Investigation Section's purpose was to locate the taxpayer and their assets. The purpose of the Research Section was to answer incoming calls and correspondence, and to adjust accounts as appropriate. Each Section was headed by a Section Chief, under whom were Unit Chiefs. Reporting to the Unit Chiefs were 8–10 employees, who comprised the bulk of the ACS staff.

The employee level positions found within the ACS were classified with respect to the Function within which they were assigned. An individual assigned to a specific job title within one function was not routinely assigned to any other function. Each function comprised a group of individuals whose focus was one aspect of the entire ACS process.

Within the Investigation Unit, the principle employee was identified as an *Account Investigator.* The person holding this position was responsible for locating taxpayers and their assets. He or she was also responsible for recommending what levies be placed on taxpayer assets and following up on those recommendations. To accomplish this, the Account Investigator contacted third parties (e.g., banks, employers, neighbours, state government agencies, the US Postal Service, etc.), and researched such published data sources as telephone directories, in order to locate taxpayers and their assets.

Within the Research Function, the principle employee was called a *Research Analyst*. The Research Analyst was responsible for analyzing and resolving tax processing problems, responding to a variety of incoming telephone taxpayer contacts, adjusting taxpayer accounts, providing information and recommending enforcement action on all types of individual and business accounts.

Within the Contact Function, the principle employee was indentified as an *Account Representative*. The role of this position was to make direct telephone contact with the taxpayer or their representatives using the computerized telephone system. The purpose of this contact was to secure payment of delinquent taxes or the filing of delinquent tax returns. In contacting the taxpayer or their representative the Account Representative interviewed the individual to ascertain tax liability for returns, financial status, potential levy sources, or other information necessary to estimate the amount owed or to prepare tax returns if the taxpayer had failed to do so. The Account Representative, in other words, determines the taxpayer's ability to pay delinquent accounts and the actions necessary to close the case. Such actions may include a levy on taxpayer assets, a request for adjustments to the taxpayer account, transferring the account to a Field Representative (i.e., a Revenue Officer), placing the account in a 'non-collectable' status, referring the case for investigation of possible tax fraud, and setting up a partial payment schedule and dates.

Supporting these positions was the *Automated Collection Clerk*, whose duties included providing clerical support to Account Investigators, Account Representatives and Research Analysts. This support amounted to inputting and retrieving information from the ACS databases, sorting mail, typing forms, preparing correspondence and maintaining files.

Another supporting position, essential to accomplishing the goals of ACS, was that of *Computer Specialist*. The Computer Specialist, as the name implies, was responsible for all activities necessary to maintain the database ans telephone dialing systems. This included programming, system updates, file inputs and data security.

4. *The work structure of the ACS*

In practice, the Automated Collection System reflected many of the activities found in the Collection Office Function. The use of computers to maintain records, dial telephone numbers and issue notices provided the major structural changes which triggered a vast array of procedural and environmental effects.

In the process, tapes of delinquent account (TDAs and TDIs) information were forwarded from the Regional Service Center after the fourth delinquency notice was mailed. This information was entered into the database of the ACS call site. Those accounts which contained complete taxpayer information (e.g., location, phone number, levy sources, employer, etc.) were routed to the Contact Section. Telephone contact with the taxpayer was made in turn, depending on the priority of the case. Such contact priorities were assigned by a computer routine which evaluated the age of the case and the dollar amounts involved. The older the case and the greater the amounts involved, the higher the priority of the case.

Cases in the Contact Function were routed automatically to the workstation requesting a case. This request was initiated by the Account Representative pressing a 'next case' button on the keyboard. The computer displayed the case information on the screen of the employee. In all, four screens of information per case were required. The prototype of the first screen displayed to the Account Representative is illustrated below, Figure 5-3.

Figure 5-3. The prototype of the first screen display for the Account Representative, 'entity data / summary information'.

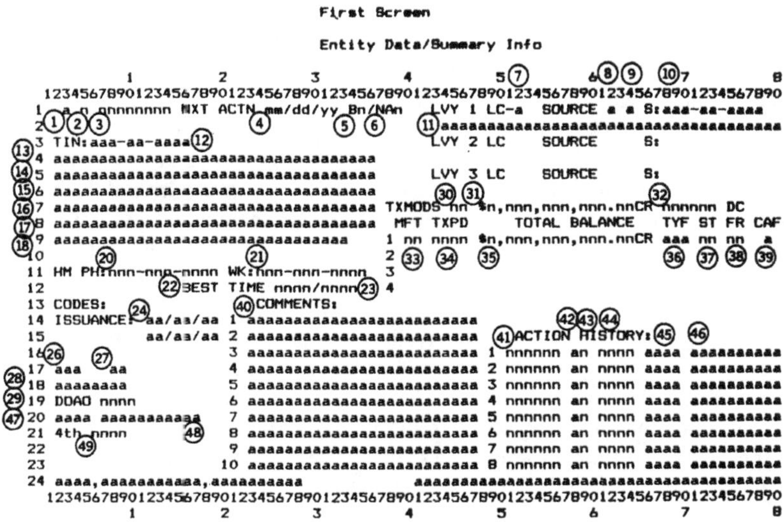

The Account Representative reviewed the case information and pressed the 'call' key. The phone system automatically routed the call through the least expensive method. Upon contact with the taxpayer of their representative, the Account Representative attempted to persuade the taxpayer to file a delinquent return or pay any amounts due. The Account Representative also tried to obtain information regarding the assets of the taxpayer, employer, address changes, or other information concerning his or her ability to pay or file. If, in the judgment of the Representative, the taxpayer was unable to make full payment, a schedule of partial payments was set up.

If the taxpayer was unwilling to agree to payments, the Account Representative could refer the case to the Field Collection Function. The Field Collection Function was staffed by Revenue Officers who attempted to impound such physical assets of the taxpayer as automobiles, equipment, machinery, etc. The Account Representative could also issue a levy notice on the taxpayer's assets. This process was accomplished automatically by means of keyboard commands which caused the issuance of a notice of levy document to the appropriate organization.

All actions made by the Account Representative or information which was obtained were entered into the file by means of annotated, standardized codes and abbreviations. The goal for an experienced Account Representative was to complete the telephone contact, discussion, and resolution within a six minute per contact time-frame. During the conversation with the taxpayer, the Representative was to simultaneously enter account information and action codes.

A failure to contact caused the computer program to automatically reschedule the call for contact at a later time.

All phone contacts were subject to monitoring by the Section Chief without the knowledge of the Account Representative. The workstation of each Section Chief had the capability of providing the same case information screens seen by the Account Representative. The Section Chief could thus monitor the phone conversation while viewing the information and actions taken by the Account Representative. The Chief also had the capability of communicating with the Account Representative during the progress of the call without the taxpayer's awareness. Either the Account Representative or the Chief could signal each other by means of an electronic tone during the telephone call. By means of an electronic switch, they could converse with each other while the taxpayer was put on hold. This procedure was used when the Account Representative had a question concerning the case or when the Chief wanted to make a comment on the actions taken.

In practice, the use of call monitoring varied a great deal. A newer employee would be monitored more frequently than a more mature employee. Employees having difficulty would be monitored more frequently than those having little difficulty. Regulations called for the monitoring of employee telephone contact twice a week, each monitoring period to last between 20 minutes and 1 hour. Typically, four phone conversations were monitored in each of these sessions. A record and evaluation of the call monitoring was placed in the employee's personnel file.

The managers in ACS call sites had available to them a variety of different indices of employee performance that did not exist under the Collection Office Function. The new computer system would, automatically, compile data on telephone time (i.e., length of time the Contact employee spent in conversation with the taxpayer), screen time (i.e., length of time the employee spent reviewing data screens), and non-connect time (breaks, lunch periods). As a result, managers spent considerably more time at their desk and terminal than they had under the Collection Office Function.

The availability of quantitative information on system operations was the result of a Managerial Information System (MIS) built into the ACS. Under the Collection Office Function, reports were prepared manually. Frequently, these reports were retrospective summaries and covered a time-frame of a month or more. Full-scale operational reviews were typically done on a yearly basis, and were time-consuming and costly means of evaluating COF management and operations.

Since the computer, under the ACS, provided real-time quantitative information, the operation and productivity of the single individual, work group, section, or entire call site could be assessed in a much more precise and timely fashion. Such indices as average number of case closings, average call time, dollars collected, size of case inventory, staff time, cost of collection per thousand dollars, etc. were commonly generated. Call site summaries were developed for comparative assessment.

This increased emphasis on professionalism and accountability was reflected in what became the slogan of the ACS, "Firm but Fair". The collection of outstanding debts or tax returns from both individuals and businesses was to be treated with a firm hand. This firm hand philosophy extended, in many cases, to the treatment of ACS employees as well.

5. *The transition from COF to ACS*

The Automated Collection System represented an evolution of the Collection Office Function into the electronic age. The process did not happen overnight and required months of planning for the transition from one system to another. With hindsight, technical matters could have been planned more accurately than administrative concerns. A great number of areas existed in which little or no precise planning could be accomplished due to the large number of unknown factors. In such situations, successful implementation was determined by the degree of flexibility in the system that was possible. In some situations, the ability to plan successfully was influenced by political and personality aspects more than technical or cost considerations. A more detailed description of the planning process can be found in Chapter 3, The Politics of Automation.

Two specific areas required a great deal of attention from the system planners: case transfer and personnel redeployment. The transfer of case load operations was accomplished by requiring all incoming cases after a specific date to be entered in a database rather than paper format. The actual dates on which this was to occur varied from site to site due to the different start-up dates of the entire ACS phase-in. Following the transfer date, no residual COF operations were to be accomplished. Some operations which were part of the COF were transferred to other Divisions along with funding for their operation. For example, walk-in taxpayers would be handled by a different group of employees, called Taxpayer Service Representatives.

COF case inventories were to be liquidated as soon as possible following the start-up of the ACS. Generally, work transfers were to be accomplished either at the inception of ACS or during the 4½-month period specified for the phase out of existing COFs. In essence, this process involved a high degree of co-ordination among different functions. The results of this process, for the most part, were very satisfactory. As ACS sites came on line, they had the experience of previous sites in dealing with their phase-out operations. The first sites to convert to the ACS were specifically designed to 'pilot' the core procedures used in other sites. Needless to say, the transition period for the later sites was smoother than for the first sites to go on-line.

The transition of personnel from the Collection Office Function to the Automated Collection System proved to be much more traumatic in nature. The overriding consideration for personnel was the fact that the ACS would require many fewer employees than the COF. In all, a reduction in staffing from about 5000 to 2500 employees was anticipated. The skills required of these new positions differed in many ways from those required in the Collection Office Function. Job security is a cornerstone of federal employment. This is especially true in the IRS which has never had a significant reduction in staffing. With this as a background, the forthcoming transition period was the cause of much anxiety and extensive planning on the part of the ACS Project Team.

Each COF site and District was assigned the responsibility for establishing a policy and procedure for staffing the ACS site and dealing with COF employees. Individual ACS sites were provided with a projected staffing plan per job category. The initial start-up period for the call sites included an 18% overhead staffing allocation to deal with initially high demand and interim productivity goals, estimated learning curves, and on-the-job training requirements.

The staffing of the ACS call sites was approached in a number of ways. Collection Office Function employees were advised of the transition through use of posted announcements and a brief video-taped film which was distributed to all Districts. Vacancy announcements were circulated and employees advised to apply for the new positions. Each site, using its own selection criteria and procedure, determined the eligibility of COF employees. Those employees who chose not to apply for or were not accepted into the ACS operation were reassigned to other, similar, positions within the IRS with no loss of pay or grade. Many sites, using the long lead time for the start-up of the ACS operation, relied heavily on attrition to reduce the number of employees who were directly affected by the transition. In many cases, new employees were recruited from the local area. The availability of federal employment was advertized in local newspapers and employment bureaux. In some instances, new ACS employees were recruited from other IRS divisions.

The National Office directed and co-ordinated the various personnel activities related to the staffing of the ACS call sites. The National Office Project Team was responsible for:

Developing and implementing the redeployment system.
Developing new position descriptions and classification guidelines.
Developing critical job elements and performance standards.
Co-ordinating and negotiating with the National Treasury Employee's Union.
Developing new test selection procedures for positions in the ACS.
Reviewing residual work after ACS implementation.
Operation of inter-office placement system (redeployment).
Developing staffing guidelines for filing new positions.
Providing technical advice to various committees and regions on such decisions or requirements that they might reach.

The Regional Offices were to implement the National Office efforts, co-ordinating activities with the District Office, resolve local labor–management issues and develop regional guidelines and procedures. Left to the District Offices was the responsibility to implement national and regional directives, staff the new organization and resolve any local labor–management issues.

It would not be realistic to say that the staffing of the ACS progressed without any glitches. For the most part, however, staffing quotas for each new site were met and the redeployment of COF personnel was accomplished. The type of person employed in the ACS proved to be very different from that found in the Collection Office Function. To a large degree, this was the result of a depressed national economy at the time of transition. Many more men were recruited into the ACS. The average educational level of the new ACS employee exceeded that of the COF employee, and the new employees tended to be younger, with different attitudes towards work and job satisfaction. A more detailed account of the socio-demographic and attitudinal characteristics of these new employees is to be found in later chapters.

Training, both for managers and technicians, was a major focus of the transition. Due to the number of issues involved in this topic, training for automation will be dealt with separately in Chapter 8.

We now turn to a description of the actual human resources activities that occurred during the implementation process. These activities are described under the headings of Personnel Recruitment and Retention (Chapter 6), Human Factors (Chapter 7), Training (Chapter 8), Team Building and Resources (Chapter 9), and Organization and Management (Chapter 10).

Although the organization of the material makes it appear that these activities were separate, discrete and sequential, such was not the case. The sheer number of people and activities that were involved within such a compressed time frame sometimes resulted in a less than orderly progression of events. As will become apparent in the following chapters, the many accomplishments of the people involved in the implementation process were the single most important factor in the final success of the system.

Chapter 6
Personnel Recruitment and Retention

Introduction

Little emphasis needs to be placed on the fact that the implementation of a large-scale automation effort requires a never ending series of activities. Even at the time of writing, three years after the first pilot sites were brought on-line, several modifications and enhancements to the system are under way. The first year of operation proved to be both a trying and exciting time for those involved. The number of separate but interdependent operations that had to be performed in accordance with strict deadlines and costs was, in retrospect, staggering. Many of those activities were of a technical nature, dealing with system parameters, computer hardware and software, site construction requirements, etc. These aspects are, of course, not the topic of consideration of this book. What this chapter will focus on are the activities that were essential to the implementation from a human factors perspective. Specifically, this chapter will deal with the issues of personnel recruitment/retention, the work process, the work environment, work equipment, training, team building and such aspects of organizational administration as they related to the ACS implementation. As discussed in Chapter 4, data drawn from attitude surveys, on-site observations, individual and group interviews form the basis upon which this chapter is written.

1. Background

The bedrock upon which any organization is built is the people who staff it. Their competencies, motivations and commitment to the organization are the pivot points around which the success or failure of any program is built. So far as the Automated Collection System was concerned, the fundamental personnel-related issues were of three types. First was the necessity to recruit and hire a sufficient number of employees to fully staff each call site as it came on-line. Second was the issue of employee skills required to competently perform ACS functions. Third was the issue of maintaining appropriate staffing levels of competent employees over time. While employee training is an essential aspect of these three issues, it will be discussed as a separate topic.

During the start-up period for the ACS call sites it was necessary to select a large number of employees in a short time frame. The goal for each call site was to over-staff positions under the assumption that, from the first day of operation, a backlog of cases would need to be dealt with. This initial surge was the result of the transfer of cases from the Collection Office Function preceeding the ACS implementation.

The recruitment of a sufficient number of applicants for ACS positions took several approaches. A primary focus of these recruitment efforts was on the COF employees to join the new ACS branch. A concerted effort was made to promote transfers from this fading function. Word of mouth quickly preceeded any formal announcement of available positions. Attempts were made to familiarize current employees with the ACS operation by means of a video-tape introduction which was widely distributed among COFs.

The forthcoming change-over from a manual to an automated system was greeted with both positive attitudes and resistance to change among the COF employees. Discussions with COF employees indicated that a number of different concerns were contributing to the resistance. Foremost among these concerns were:

Concern with loss of job/status.
Increased pressure to perform more 'intellectual' work.
Demands for more productivity in a shorter time frame (i.e., pressure).
Job will be simplified, resulting in boredom.
Loss of familiar routine and established social relationships.
Fear of failure using new equipment and routines.
Loss of established career path and performance criteria.
Irritation at having to learn new skills after having spent years building up the current level of competence.
Distrust of new technology (e.g., "Can I push the wrong button and destroy the system?", "I can see words when I type them but I don't really trust in computers").

Many COF employees, of course, looked forward to the new automation. In general, they were the newer employees, less pessimistic, and with less investment in the older way of doing things. Employees who could not or would not make the transition (including those who were not selected by ACS) were transferred to other operations.

Efforts were also made to promote ACS employment within other branches of the IRS. A third source of applicants was 'new hires', i.e., people not currently employed by the IRS. While position vacancy notices, word of mouth, and promotional discussions were held to attract ACS employees from within the service, newspaper advertisements, state unemployment agency vacancy notices and current IRS employees were used to spread the word regarding the availability of positions within the new Automated Collection System.

These efforts towards recruiting new employees were generally successful. The promise of working in a modern facility with 'state of the art' equipment was enticing to COF employees who were accustomed to manual operations and less than ideal office environments. Many simply switched over from COF to ACS because the alternative, to be placed in an unfamiliar IRS operation, was unacceptable.

The difference between those newly hired to the IRS and those who had previous IRS employment was, in many ways, dramatic. As cited in the HRT final report some of the most significant differences were:

1. Sixty per cent of the newer employees were full-time permanent compared to 92% of the long-term employees.

2. Newer employees were significantly younger than long-term employees. The average age of the newer employee was 25 years compared to an average of 33 for the long-term employee.

3. Educationally, over 53% of the newer employees held a college or graduate degree, compared to 24% of the long-term employees.

4. Future career expectations of the two groups differed. Only 3% of the new employees indicated that they expected to be working in their current position within the next few years. Almost 14% of the long-term employees indicated that they would be in such a position within the next few years.

5. The newer employees also indicated different reasons for possibly leaving the ACS than the long-term employees. Pay/benefits and promotional opportunities were key ingredients for the newer employees but few long-term employees indicated that they had any plans to leave the ACS.

6. When asked "What first attracted you most to employment with ACS?", the most common answers were continued IRS employment, the new work technology and the opportunity to learn new skills. The nature of the work itself was also frequently mentioned as an attraction.

7. New hires accounted for a significant proportion of the ACS employees. Although the percentages varied by site, over 40% of the new ACS employees were drawn from outside the IRS. Indianapolis showed the fewest percentage of outside hires (13%) while Nashville had the highest (almost 60%).

8. Many of these new employees were not employed as full-time permanent staff. Again, although the figures varied from site to site, only about 60% of ACS employees were hired on such a basis. In Nashville, for example, less than 42% indicated that they were full-time and permanent. It is likely that new employees who accepted a position that was other than a permanent, full-time one were motivated by one of three conditions. First, they may have desired part-time, temporary employment. Second, they were unemployed and willing to accept conditions of employment that were less than desirable as a means of bridging the period of time until they secured a better job. Third, they hoped that the part-time, temporary appointment would lead to a more permanent, full-time appointment. Our experience indicated that the latter two conditions were, by far, the most prevalent.

As a whole, the ACS workforce was in some ways different than one might have anticipated given the entry level skills and pay. The typical employee was about 34 years of age. Almost 70% were female. Most ACS employees used their income to support only themselves or as a secondary income in a two-person household. Over 30% of these employees possessed a bachelor's or higher degree. About 13% had either a graduate degree or some graduate level courses. The relatively high rates of unemployment in the geographical areas surrounding the call site appeared to be the most significant factor in attracting a highly educated work force willing to accept a non-permanent or temporary appointment.

Taken together, these factors create a pattern that raised many issues that had the potential for a significant impact upon the ACS organization. The seeds of the future were planted in the personnel selected for employment.

A significant number of COF employees either chose not to transfer to the ACS or were not selected. Those who were hired were younger and better educated, and had fewer people dependent on their income than the typical IRS employee. Many had been hired in an other than full-time capacity. Most significantly, perhaps, their expectations of the future were high. The career opportunities within the ACS were fairly restricted, one supervisor for about 10 employees. Unlike the Collection Office Function, a career ladder path had not been formally established for upward mobility by means of other, related occupations. The contribution of these factors to the personnel retention problems was a factor in a follow-up study (to be detailed below). Sites varied tremendously from each other in terms of who and how they recruited ACS employees.

The ACS experiences, with respect to the IRS, an unusually high rate of staff turnover. On the average, the ACS turnover rates are from 30 to 40% nationwide. Some experience turnover rate as low as 5% while some exceed 100% per year. In a few cases, the labor pool of applicants has been nearly exhausted and new populations of potential employees are actively being sought to replace those who leave. It is important to note in interpreting these figures that private sector collection agencies, a very rough parallel to the ACS operation, also experience very high turnover rates. In many cases, the private sector collection agencies' rates greatly exceed those of the ACS operation.

An employment factor which might have had significant consequences for the long-term effects of personnel on the ACS was the future aspirations of these new employees. When asked what their expectations were within the next few years almost 40% of the pilot site employees expressed a belief that they would be promoted within their job series. A slightly lower percentage of employees believed that they would stay within the IRS but leave the ACS. The responses to the question "Within the next few years I would expect that:" are displayed in Table 6–1. A very high percentage of the data comparison sites at Denver and Boston expected to stay with the IRS, but not as a member of the ACS.

Table 6–1. Employees' expectation of their employment in the next four years.

Answer	% response
I will be working at the same job I now have	8.4
I will be promoted within this job series	39.8
I will stay in the IRS but leave ACS	35.0
I will be working at another state/federal job	4.2
I will be working in private industry	7.3
I will not be working	4.2
Other	1.1

Expectations of the ACS job were also measured. For the pilot sites, an average of about 75% indicated that the job was about what they expected, better, or much better than they expected. This is illustrated in Table 6–2. For the control sites, the average response in the same categories (i.e., "about what I expected, better or much better") was 68%. A parallel issue, what would most likely influence ACS employees to leave, was also asked. Employees indicated two related reasons with a far greater frequency than all others—lack of advancement opportunities and salary/benefits.

Table 6–2. Employees' assessment of their current job.

Answer	% response
Much worse than I expected	9.5
Worse than I expected	24.6
About what I expected	45.2
Better than I expected	14.6
Much better than I expected	6.1

One call site, experiencing high attrition rates, identified the problem as unrealistic expectations of the work and the future career potential within the job series. To attract as large a number of applicants as possible, recruiters had been selling the positive aspects of federal employment, working in a modern facility, the importance of the position, etc. The fact that these were entry-level positions, often requiring long stretches of sedentary time talking with sometimes hostile taxpayers, was not emphasized.

A revision of this approach attempted to present a more realistic picture of the job and the employee future that could be expected. Although not a controlled study, it was felt by those involved in the revised recruitment procedure that the type of employee entering the ACS system would have much more realistic expectations and be more likely to remain. It was also noted that, after the presentations of the 'realistic' description, the number of people who actually applied for employment dropped dramatically. This raises the issue of whether it is better to be understaffed but maintain a stable workforce or to be fully staffed with a mobile workforce. The resolution of the issue concerns the cost of employee replacement, whether the turnover is primarily extra-organizational or to other divisions within the organization, and the continued availability of an appropriate labor force.

In discussing employee turnover with ACS executives, it occurred to us that the physical withdrawal of employees from the ACS organization represented only the immediately observable act of leaving. Turnover, we argued, was occurring in many forms throughout the ACS. On a psychological level, an employee may withdraw by reacting to the work environment in an apathetic manner, exerting the least possible effort toward task accomplishment. On a behavioral level, withdrawal may be expressed as work absences, tardiness, or, in extreme situations, turnover. Both types of withdrawal have significant consequences for both the employee and the organization.

The negative consequences of staff turnover are obvious. From the viewpoint of the organization, employees who put in a minimal effort are less productive than those who are more involved in their work. Those who leave represent a loss to the organization of costly recruitment and training expenses in addition to the loss of a valuable experience base. They leave behind them a vacancy which frequently requires additional effort on the part of their co-workers and supervisor to 'pick up the slack'. Such a turnstile effect tends to lower group morale and promotes a continual cycle of turnover.

From the individual point of view, withdrawal from the organization, either on a psychological level or a behavioral level, may also have negative consequences. Although the individual commonly blames the organization, he or she experiences a sense of personal failure in connection with withdrawal. The experienced inability to cope with the situation, disappointment with lack of success, guilt, and the stress associated with major job or career change also contribute to the sense of disappointment.

On the positive side, turnover can be viewed as part of the selection process. Those employees who do not 'fit' tend to leave, new employees bring a fresh perspective and energy to the organization. So far as the employee is concerned, leaving is a means of starting over with an increased knowledge base. It is also a means of career progression and finding one's niche in life.

In essence, turnover can be positive, negative or neutral. Positive turnover is the result of promotion, and career ladder moves. Both of these processes result in retaining skilled employees, a valuable resource for any organization. Another form of positive turnover is the loss of unproductive employees. This form of turnover can be the result of firing or self-selection out of the organization. In either case, those employees who are willing or unable to fit into the organization leave.

Retirement, although it usually involves the loss of skilled and experienced employees, is an example of neutral attrition. It is not the result of individual or organizational forces in the same sense as positive or negative turnover.

There exists only one form of purely negative turnover—the loss of competent employees who leave the organization entirely. These are the employees that the organization would like to retain but who leave of their own volition.

Some base-line turnover of this latter type is inevitable due to changing lifestyles, careers, geographical movement, etc. Another factor affecting this base-rate is the occupational category of the employee. Some occupational groups tend to be more stable than others. In general in the USA, employees in the service industries have an average tenure rate of 4.1 years compared to an average of 8.1 for executive, administrative and managerial personnel. Base rates will also vary as a function of local economic conditions. Organizations located in a high growth area will face stiffer competition for human resources than those organizations located in a stable or decreasing economic environment.

Two points need to be stressed. First, staff turnover is not necessarily bad; it can benefit both the organization and the individual. Second, depending on the occupational category, a base-line turnover rate can be anticipated. What is demonstrably expensive both in human and organizational terms is negative turnover which significantly and continually exceeds the occupational base rates.

Such a condition, excessive turnover of competent employees, was considered to exist within the Automated Collection System (ACS) of the Internal Revenue Service. To investigate the nature and causes of ACS employee turnover, a study was jointly designed and conducted by the Human Resources Technology Staff and the Collection Division. This study focused exclusively on identifying the factors that might relate to employee turnover. It was administered completely independently of the ACS "What's Your Opinion?" study described previously. The goal of the study was to identify the attitudinal and socio-demographic factors, if any, which could be used to predict employee turnover.

2. Research methodology

A self-administered survey was designed to collect employee data in four issue areas: (1) socio-demographic background; (2) satisfaction with specific conditions of the work and work environment; (3) general job satisfaction; and (4) turnover likelihood issues (see Appendix 2). In all, 109 questions were asked. The survey took ten minutes for the average individual to complete. Along with the survey, participants were given a stamped, addressed postcard which they were to keep until such time as they decided to leave the ACS. At the time of that decision, they were instructed to complete the two questions on the postcard ("Why did you leave the ACS?" and "Where did you go?") and mail the card back to the researchers.

In addition to the socio-demographic and attitudinal data, each employee and supervisor who completed the questionnaire was rated on performance by his or her supervisor. Three performance criteria were used: quality, quantity and dependability of performance. Each criterion was rated on a ten-point scale. The questionnaires were collected from three pairs of ACS sites. Two sites were identified as having a relatively low turnover rate, two sites as having a moderate rate, and two additional sites as having a high rate. The identification of these sites as representing high, moderate, or low rates of employee turnover was based on the judgment of the Collection Division staff rather than a statistically based decision. This approach was necessary due to the lack of current or consistent data among ACS calls sites with respect to staff turnover records.

Data collection occurred during the period from October 1985 to April 1986, and involved 506 participants from all levels in the ACS hierarchy, excluding only the position of Branch Chief. A copy of the questionnaire was left at the site for those not present during the administration period. Instructions for completing and mailing the questionnaire were also provided.

3. Data analysis procedures

A two-stage analysis procedure was used. First, summary statistics and frequencies were compiled for each of the three pairs of sites (i.e., high, medium and low rates). Second, correlation analysis among selected variables was conducted. The correlation analysis consisted of examining the relationship among all variables with respect to the two dependent variables: intention to leave the ACS and intention to leave the IRS. All analyses excluded non-responses; that is, reported percentages are based only on the number of valid responses to that question and not the total number of respondents in the survey.

Two significant lines of inquiry were pursued in this study. The first tried to identify those factors which differentiated high, moderate and low turnover rate sites. The second focused on the relationships among attitudinal and socio-demographic factors in predicting turnover. Employee intentions regarding staying or leaving have been identified in the literature (e.g., Mobley, 1982) as the single best predictor of actual turnover.

Due to a very low response rate, no attempt was made to include data from the self-mailed postcard. In all, a total of 12 postcards were received, and of this number only two contained any usable information.

4. Results: summary statistics

The responses to the questionnaire were analyzed using a statistical software program.

Performance differences

The evaluation of employee performance differed among low, middle and high sites. The employees at sites exhibiting a low turnover rate had a significantly higher average rating on the three rating factors (i.e., quality, quantity and dependability of performance). On a scale from 1 (extremely poor) to 10 (excellent), low turnover site employees averaged 7.3, compared with 6.7 for moderate and 6.9 for high turnover sites. Especially with regard to ratings of performance dependability, the low site employees were judged as being more competent than the other sites.

Socio-demographic differences

On the average, employees at high turnover sites tend to be younger than at low turnover sites. Moderate turnover sites, however, have a significantly older population than either the high or low sites. The older the employee, the more years of IRS experience the employee reported. Being older, low turnover site employees tended to have more people dependent upon their salary (either fully or partially) than high turnover sites.

Low turnover sites have a higher proportion of married and better educated employees than high turnover sites. Moderate sites show the highest percentage of married employees but they are less educated than at the other sites. About 41% of the employees at low turnover sites are college graduates or have a graduate degree compared to 31% at the high sites.

The ratio of female to male employees is significantly lower (1.7 females for every male) at low turnover sites than at moderate (2.3) and high sites (2.8).

Low turnover site employees tend to commute fewer miles than high site employees, but moderate site personnel commute even less. However, the greater the proportion of employees that commute alone in a car, the briefer the commuting time and the lower the turnover rate.

Employment status differences

No significant differences among sites were noted in terms of how they first heard of the ACS position. At all sites, about 30% heard through a vacancy notice, and about 23% read a newspaper advertisement.

Significantly, the low turnover site employees reported a longer period of time in grade than high turnover sites (18.5 months *vs* 12.3 months). Moderate turnover sites reported the highest average time in grade, almost 19 months.

The type of employment status also varied among sites. Fifty-three per cent of the low turnover site employees indicated that they were full-time permanent staff. This compares with 60% in the same category at high turnover sites. A significantly higher proportion of moderate turnover site employees (almost 80%) were full-time permanent.

On the average, low turnover site employees reported that their IRS income accounted for 66% of their total income. Spousal income and second jobs accounted for the remainder. Moderate and high turnover sites reported higher proportions (67% and 74%, respectively).

No differences among sites were noted for number of prior jobs (IRS or non-IRS), similarity to previous jobs, or applicability of education or training to their current position.

Satisfaction/dissatisfaction with specific work characteristics

This part of the questionnaire required employees to indicate, from a list of 30 work characteristics, those which they considered most and least important to them.

Regardless of turnover rates, all sites reported that salary is the most important factor in their work, followed by unanimous agreement that advancement potential was the second most important factor. Although in slightly different order, all sites agree that work that provides an opportunity to learn new skills is a significant factor. Both the low and high turnover sites agree that work in which employees are treated fairly is a significant factor in job satisfaction. Finally, employees at both the high and middle turnover sites identified working for a supervisor that they respect as important to their job.

In terms of aspects of the work that are least important, there is less uniformity among sites. Both the middle and high turnover sites agree that working for a small organization is important although the definition of organization was not made in the survey. All three types of sites agreed that work that allows outside employment and working in a high prestige job were also very unimportant. Middle and low turnover sites agree that routine and structured work is unimportant.

General job satisfaction

Questions in the general job satisfaction section of the questionnaire dealt with broad aspects of the work and the employees career. Of the seven questions raised in this section, no significant differences or patterns in average response among the three sets of call sites were noted.

Aspects of turnover

This section of the questionnaire asked about the self-reported likelihood of leaving, reasons for leaving and destination.

In answer to the question of the likelihood of leaving the Automated Collection System within the next few years, the responses for the low, moderate and high turnover sites are illustrated below (Table 6–3). H, M and L represent the average high, middle and low attrition sites, respectively.

Table 6–3. Probability of leaving the IRS, by site, in per cent.

	Site		
Probability	L	M	H
Better than 50%	35.0	27.8	40.1
50%	27.5	30.1	29.7
Less than 50%	37.5	42.0	30.2

The same relationship among the percentages for all three sites appeared in response to the question addressing the likelihood of leaving the IRS although the averages were higher for all three groups, indicating a greater probability of leaving the IRS entirely than leaving the ACS. In a practical sense, the average likelihood of leaving for the low and moderate sites was considered the same (i.e., about a 50% chance). The high attrition site did, in contrast, show a marked increase in the likelihood of leaving both the ACS and the IRS.

These positions were reversed when employees were asked the likelihood of leaving the IRS within the next year. Here, an average of 45% of the low turnover sites indicated that they would leave the IRS compared to an average of a 55% chance for the high site. The medium turnover site reported a 43% average.

All three turnover sites, when asked to indicate their feelings about changing their job, exhibited the same attitude. The most common response, about 50% of the total for each site, was "I am not eager to change my job, but would do so for a better job".

The similarity in response among the three sites was also evident in response to reasons for leaving their job. In identical rank order for each site, the top four rated reasons for leaving cited were:

More pay/promotion.
Future career opportunities.
More interesting/challenging work.
Poor management practices/policies.

5. *Results: relationships among items*

In addition to the summary description of each question asked of ACS employees presented above, the relationships among responses to different questions were analyzed. More specifically, a series of analyses were conducted to determine which factors or combination of factors could be used to predict intentions to leave the ACS and the IRS.

Consistently, throughout the analysis of relationships in each of the three groups, the predictability of employee answers regarding the probability of leaving the ACS was negligible. None of the items measuring employee background performance, job satisfaction or work/job preferences showed a predictable pattern when it came to the decision to leave the ACS.

Analysis of the relationships among questions to predict intention to leave the IRS, however, did show some significant relationships. So far as the low turnover sites are concerned, employee ratings on 'dependability of performance' correlated significantly with the intention to leave the IRS: the higher the dependability rating, the lower the likelihood that the individual would leave the IRS. Neither the middle nor the high turnover sites showed this specific relationship. The middle sites alone showed a strong relationship between satisfaction with overall career progress and intention to leave the service: the less satisfied with career progress, the greater the likelihood of leaving.

The one consistent pattern among all three sets of sites was the relationship between overall job satisfaction and intentions to leave the IRS: generally, the less satisfied, the higher the probability of leaving. While this may seem self-evident, the factors which influence overall job satisfaction as a predictor of IRS turnover are not. On the average, the level of job satisfaction is about the same in all three sites, about half way between 'satisfied' and 'dissatisfied'.

To identify those factors, if any, which correlate with job satisfaction, a secondary analysis for each of the three attrition sites was accomplished. For the low attrition sites, performance feedback proved critical to employee satisfaction. In the moderate attrition sites, working with the right equipment and procedures correlated with job satisfaction. In contrast, the high attrition sites showed no consistent pattern between other factors and job satisfaction.

6. *Study summary*

The factors which clearly differentiate the high, middle and low turnover sites in a straightforward fashion are few. To summarize the foregoing description, the following patterns among all three sites were evident in the data analysis:

1. The lower the ratio of females to males, the lower the turnover rate.
2. The higher the percentage of college graduates, the lower the turnover rate.
3. Sites where employees depend on public transportation for commuting have a higher turnover rate than those whose employees drive alone to work in their own car.
4. The longer the commuting time, the higher the turnover rate.
5. Salary and advancement opportunities are rated the most important factors in work followed by work that provides the opportunity to learn a new skill.
6. The higher the overall job satisfaction, the lower the likelihood of leaving.

Since site selection was somewhat arbitrary, comparing data from the high and low sites alone might provide a more realistic and straightforward analysis. Thus, comparing only the high and low sites, the following implications can be drawn from the data in addition to the overall trends among the three sites previously discussed:

1. The higher the rating of employees (i.e., quality, quantity and dependability of performance), the lower the turnover rate.
2. The older, more experienced, married employees, tend to leave at a lower rate than single, less educated, employees.
3. The greater the dependence on IRS salary for total income, the greater the likelihood of leaving.
4. Low turnover site employees indicate a higher likelihood of leaving the ACS but a lower chance of leaving the IRS than is the case with high turnover site employees.

7. Recommendations on recruitment and staff turnover

The issues raised above are but a few of those that need to be addressed. Based on our own experiences, the following actions may reduce the negative impact of traditional personnel practices in staffing the automated office:

1. Perform a job analysis to identify the abilities and capabilities that are required to perform the job competently. Selection should be made only on these criteria. The approach of getting 'the best employee for the job' should focus on these criteria.
2. Establish a career ladder for growth within each occupation.
3. Use current employees as a primary source of obtaining applicants.
4. Require current employees and their manager to participate in the selection interview process.
5. Promote the movement of competent employees to other branches and divisions. This 'positive employee flow' is based on the assumption that all employees are temporarily assigned to their current job.
6. Estimate the cost of recruiting and training the typical employee. How long does that person have to be productive in their job before the organization has received a return on its investment? This length of time at a given level of productivity should provide a standard against which job tenure is judged.
7. Employees should be hired with a placement philosophy, not simply a selection/non-selection attitude. Failure in one position should be an indication of poor placement rather than a failure of the selection process. With few exceptions, the specific interests and capabilities of a given individual can be useful in some aspect of the organization's operations. One way to promote this is by the establishment of cross-functional advisory committees of peers which might recommend work activities, options, educational credentials and career progress milestones that might be beneficial to the organization and individual involved.
8. Allow as much flexibility in work arrangements as possible. To attract and retain a highly competent and motivated workforce allow work schedules to vary to meet individual needs. The needs of working mothers are of special concern. It is better to have two motivated part-time workers than one full-time mediocre employee.
9. Recruit realistically. Employees resent being tricked into taking a job which doesn't turn out as promised or implied. It is much more beneficial to be honest in describing both the positive and the negative aspects of the position. The increase in rejection of job offers that results is more than compensated for by the reduced post-selection turnover that occurs after thousands of dollars and months of time have been invested in bringing the employee up to a satisfactory level of competence only to have them leave with all the skills you have provided him or her.
10. Part of turnover comes from retirement. Unfortunately, retiring employees are often some of the most valuable employees due to their experience. The period preceeding retirement has, traditionally, become a marking time period of decreasing involvement. One way to reduce this effect is to provide alternatives to the traditional retirement system. Establish a senior employee program which could act as a trouble shooting consultancy or an on-the-job training scheme in different functions. The work schedules of these senior employees might vary from full-time to one day a week or might incorporate a flexitime schedule. The senior employee would have the advantage of a higher pay level than full retirement, the flexibility of a part-time work schedule and the psychological benefit of useful activity. The organization would benefit from their experience and reduction in non-productive retirement pay.

These recommendations were presented to the Collection Division at the conclusion of the study. Although it was agreed that these were viable methods to improve the recruitment and retention of employees, the decision to implement any action was left to the local district. To the best of our knowledge, only one site has attempted any significant effort. That site has developed a recruitment program based on an accurate description of the position and its advancement opportunities. They report that the self-selection out has increased but those that remain are much more likely to stay.

Chapter 7
Human Factors

Introduction

The easiest aspect of an automated system to ignore is the human factors component. Under pressure to meet deadline and cost constraints, the time, expertise and expense required to assure that the automated systems and the work environment are compatible with the needs of the workforce is frequently given a low priority. It is easy to point to measurable effects of well-planned equipment and software. The effects of an ergonomically well-designed work environment are less demonstrable. Employees, it is often argued, are less expensive, easier to replace and more adaptable than other components of the system. What planners misunderstand in this logic is that they are focusing on the near-term outcomes rather than the long-term consequences.

Employees are, with very few exceptions, by far the most costly aspect of any organization. To the direct cost of salary and benefits must be added the costs of recruiting, selecting, training and placement, sick leave, annual leave, retirement plans, etc. If it was common practice for organizations to compute the actual direct and indirect costs of personnel during their tenure, the costs incurred by addressing ergonomic concerns and job satisfaction issues would seem small in comparison.

1. Ergonomic considerations

Planning for the Automated Collection System did attempt to incorporate some consideration of ergonomics. Since the majority of the equipment was purchased 'off the shelf' and had been in widespread use in the private sector for several years, it was assumed by the IRS project staff that it had already been subject to acceptability testing.

The major ergonomic consideration of the equipment was its adjustability. For example, the work table, the chair and the video display terminal (VDT) incorporated limited degrees of swivel and tilt movement. The terminal keyboard was movable and included optional wrist supports. Similarly, the desk included an optional footrest and could be adjusted for height, as could the operator's chair.

The major ergonomic problems that arose during and following the implementation were, in somewhat piecemeal fashion, eventually corrected or retrofitted. Overall, survey results indicated that employees were generally satisfied with the ergonomic aspects of the hardware and software. Using a four-point scale of satisfaction—1 = very unsatisfied, 2 = unsatisfied, 3 = satisfied, 4 = very satisfied—Table 7-1 summarizes the reaction of employees to the ergonomic aspects of their job.

Each aspect cited above represents a weighted composite of several questionnaire items. Since the control group sites were not equipped with similar technology, no such comparisons were appropriate.

Table 7-1. Average satisfaction with the work environment.

Aspect	During implementation	After implementation
Work desk	2.78	2.80
Work chair	2.92	2.86
Storage area	2.79	2.74
VDT	2.99	2.89
keyboard	3.06	3.03
software	2.80	2.77

Interpreting the results shows a clear pattern of moderate levels of satisfaction with the new technology. Broken down by prior position, those ACS employees who came from the old COF showed a significantly higher level of satisfaction compared to the newer employees. Those who had experienced the frustrations under the manual system were probably more appreciative of the automated systems. Table 7-1 shows that there is a slight, but consistent, trend for satisfaction levels, among all aspects (with the exception of the work desk), to decrease over time. This may be attributed to three reactions: the novelty (Hawthorne effects) tends to wear off, subtle problems show up only after extended use, or minor irritants, ignored at first, tend to become more important over time.

Several aspects of the technology did cause concern among employees. Generally, these concerns could be classified as having their origin in the characteristics of the ACS equipment or being attributable to the software design.

Prominent among the equipment issues was screen glare. The surface of the CRT tended to reflect the overhead lighting, which was flourescent tubes. Several remedies for this were proposed by the HRT staff, including relocating the workstation, retrofitting the flourescent fixtures with diffractors to diffuse their illumination, and purchasing glare filters which could be placed over the face of the CRT. Most sites employed at least one or a combination of these remedies.

Some of the problems relating to the equipment were due to a lack of knowledge regarding their proper use. This was especially true of the operator chairs. Initially, seating problems were the result of poor equipment manufacture. The vendor was contacted and this problem was quickly addressed. The inability of some operators to properly adjust the equipment (i.e., chairs, tables, VDT) often resulted in complaints of headaches and back pains. Instructions that were supplied by the manufacturer were often lost or ignored. The HRT staff designed and distributed a brief pamphlet called *Using Your ACS Workstation* which provided directions on how to adjust the equipment to fit the employees' personal preferences.

Some of the earliest problems that arose concerned the software. One of these concerned 'page' time. Taxpayer information was contained on a total of four screens. During the start-up phase, the time required to move back and forth through these pages was sometimes excessive. Delays of several minutes were not uncommon. With the taxpayer on the telephone line, Contact Section employees were often uncomfortable waiting for the computer response. Part of this problem resulted from simultaneously using the main computers for training and operation—too much demand on the system. Another part of the problem was caused by inefficient software which created and recreated the pages every time they were called up. The solution to this problem was simple, if expensive. Additional memory units were added to the main computer. Except during periods of extreme demand, no complaints of slow page time have been noted since.

Another software-related problem initially resulted in a great deal of operator frustration. Part of the ACS operation required Contact employees to indicate what actions were taken as a result of the call. These actions were being recorded by technicians in a small 'Comments' field on the first screen. Abbreviated action codes, due to space restrictions, were employed for this purpose. Unfortunately, many sites tended to invent their own coding scheme with little co-ordination among sites. Even with these 'brevity' codes, the comments field tended to fill up quickly. The technicians frequently had to make the decision as to what information to drop or summarize when they needed to add additional comments. These problems were eventually resolved by establishing common codes and by increasing the amount of space available for such comments.

An additional equipment problem arose with regard to the telephone headsets used by operators. The headsets originally purchased included an ear-piece that was inserted into the ear. Although each operator was issued with their own headset, several reported ear infections with extended use of the headset. The solution to the problem was to make available 'walkman' type headsets whose ear piece rested on the external portion of the ear. Operators were given their choice of sets.

One of the most debated issues during the initial phase of implementation concerned the need for typing skills for Contact personnel. The original plan (and position description) did not require such skills of the operators. In practice, many operators felt that such skills were required since entering information was essentially a keyboard operation. The issue appeared to resolve itself over time since it became apparent that the amount and type of keyboarding skills required of operators were very different from those of a typical typist. Employees quickly adapted to the job requirements.

A more serious concern also related to data collection and entry by Contact Section operators. The intention of the system planners was to have the Contact operators enter data or modify existing information at the same time as they were in conversation with the taxpayer. This procedure would require the almost simultaneous performance of two tasks. Many operators rebelled against this procedure. In practice, many would keep paper and pencil notes during the telephone conversation. Following the call, they would enter the information into the system records. Since this was to be a 'paperless' system, many supervisors reacted negatively to this procedure. One site went so far as to prevent employees bringing pads of paper into the call site. The issue was compounded by the fact that the time goal for each call was six minutes. Since many employees were not taking the time to enter information into the computer records during the actual call period they were, in effect, finding a way around the time limits. The solution to this issue was two-fold. Some sites have ignored the restriction on note taking. At others, many employees have become more adept at entering information during the conversation, not following it.

2. *The workstation environment*

Particularly annoying to employees were restrictions on their ability to personalize their workstations. The design of the workstations, for most sites, followed that of an X, each of the four slices containing a complete workstation. The dividers between the stations were 5 foot high (1.6m) fabric covered panels. No personal articles—pictures, calendars, awards, drawings, etc.—were permitted. Two reasons were given for this. First, individuals were not assigned to any particular workstation. They were assigned to any open station. It was felt that the personalization of workstations would promote the feeling of ownership of specific stations. Second, was sound reduction. Pictures, photographs, calendars, etc. placed on the dividers would, it was believed, act as sound reflectors and degrade the sound deadening properties of the dividers. All personal effects were to be kept in mobile storage units which were assigned to individual employees. These could be wheeled around to the assigned workstation by the employee at the beginning of the shift.

Almost equally annoying to many employees were restrictions on eating, drinking or smoking at the workstation. Some sites were not uniform in their prohibition of these activities. Some sites allowed smoking but not eating or drinking. Some sites allowed drinking but not eating. Employees were, of course, aware of these discrepancies. Those sites which allowed smoking had installed a high efficiency air circulation system which eliminated smoke from the air. Other sites soon followed this example and all sites now allow smoking at the workstation. It was more difficult to resolve the issue of drinking at the workstation. The fear was that spilled drinks would damage the equipment. As a compromise, some sites allowed 'spill-proof' coffee cups. The situation proved to be particularly vexing to Contact employees whose job required them to talk on the phone for extended periods of time. Recently, a National Office memorandum was issued allowing drinking at the workstation. On reflection, the issue of drinking at the workstation disappeared of its own accord. No incidents of spillage damage to equipment were ever reported.

Satisfaction with other aspects of the work environment were also measured. Compared to the non-automated COF control groups, employees in the ACS function were much more satisfied with such factors as temperature and humidity levels, noise levels, lighting, supplies, orderliness of the work area, etc. Using the same four-point scale as previously described, the average level of satisfaction for the pilot sites with the work environment was 2.65 (third administration of the questionnaire) compared to an average of 1.96 for the control sites.

3. *Changes in job satisfaction*

Another set of employee human factors issues may be defined in terms of changes in the level of satisfaction with the job and the work process over the course of the implementation. These changes occurred during the period of time from the second to the third administration of the attitude survey. Questionnaire results indicate that satisfaction levels decreased for the following aspects of the work process:

Total amount of the workload.
The speed or pace at which they were expected to work.
The emphasis on productivity.
The amount of challenge in their work.
The amount of interruptions.
The amount of control over their work.
The availability of help.
The variety of things that they did.

In contrast, the satisfaction levels of several other aspects of the work process increased, specifically:

The changes in the workload (peaks and valleys).
The amount of feedback and help from co-workers.
The ease of housekeeping chores.

The amount of feedback and help from their supervisor.
The amount of paperwork.

In a very general sense, it might be concluded that satisfaction levels with the technical, labor-saving aspects of the system tended to increase but satisfaction with the psychological aspects of the work tended to decrease.

Satisfaction with the work itself and with the job as a whole were relatively high. Using the four-point scale again, a composite measure of overall satisfaction was fairly high although it decreased from an average of 3.1 to 3.0 from the first administration of the survey to the third. These findings should be compared with the overall job satisfaction levels at the control sites, which remained stable at around 2.6.

Several organizational and administrative issues which had a direct impact on employees' satisfaction with the job will be discussed in Chapter 9. Of particular concern here are the issues of monitoring, physical restriction, and case follow-through.

Managers were required to monitor employee telephone contact several times a week. The manager had the ability to listen in on telephone conversations without employee awareness. The manager, in addition, had the technical ability to 'barge in' on calls in progress when he or she felt the employee needed correction or guidance. The taxpayer remained unaware of the discussion between supervisor and employee. In many situations, Contact employees felt that this procedure was used in an unnecessarily punitive fashion since the monitoring formed a basis of their performance appraisal. The employee also had the ability to signal the supervisor that he or she should monitor the telephone conversation if an issue arose on which the employee felt that he or she needed advice.

The second issue which appeared to directly affect the job satisfaction of employees was the physical restriction of the workstation. During telephone contact periods, employees were instructed to remain sedentary at their workstations until appropriate break periods. In some cases, employees were required to raise their hands and sign off their terminals in order to take a break. In derision, some employees referred to the telephone headset as their 'umbilical cord'. Many found the emphasis on keystrokes, screen time, average contact time and strict adherence to break periods demeaning.

The third issue concerned task completion. This was composed of two distinct parts. Except in those cases when a satisfactory agreement was made with the taxpayer on the first call, the employee rarely learned of its final disposition. Follow-up calls, if required, were done by the next person who happened to push the 'next case' button. Even when one-call agreements were reached, the Contact employee had no way of knowing whether the taxpayer followed through with their agreement. In other words, the employee almost never received feedback of their success or failure rate since cases were not assigned to specific individuals. This situation was compounded when taxpayers called back in response to an ACS contact. Call backs were assigned to the next available operator who was usually unfamiliar with the case and unprepared to deal with sometimes hostile taxpayers. As a result, most call site personnel would much rather make outgoing than take incoming calls.

Employees tended to be more satisfied with working for the IRS as an organization than they were with working for the ACS. The highest levels of satisfaction recorded were with respect to the spirit of teamwork that existed between co-workers.

In summary, it would appear that employees were fairly satisfied with the technical aspects of automation and their work environment. While a number of ergonomic problems did arise, these were addressed through retrofitting, adaption of the individuals, or enhancements to the system. The major human factors issues that were not immediately addressed concerned the psychological impact of the work on the employees. It is only recently (1986) that attempts are being made to address these. The focus of these attempts is a 'team' concept for the ACS sites. ACS is experimenting with the replacement of separate functions (Contact, Research and Investigation) in favor of a work team whose members will assume all three functions. The work team will be responsible for the entire sequence of activities rather than assignment to one part of the task. Hopefully, this approach will alleviate some of the negative psychological consequences that were evidenced in the study.

Chapter 8
Training

Mary Ann Ruth

Preface

In the minds of most managers and executives, 'human resources management' refers entirely to personnel and training activities. Personnel activities include such things as preparing position descriptions, setting compensation, staffing, etc. Many of the ACS actions with regard to these personnel aspects have been discussed in Chapter 7.

This chapter describes the other half of the human resources coin, ACS training. In a very broad sense, training encompassed three areas:

1. Technical training about the equipment and systems.
2. Tax law, rules and procedures.
3. Human relations and management training.

A description and interpretation of the training activities is presented here by Mary Ann Ruth, the staff member of the Human Resources Technology Group most responsible for designing and implementing the human relations and management training cited above.

Going beyond the stereotype of a two-dimensional human resources contribution, some of the following chapters deal with employee and executive development and motivation.

Introduction

Training, as usually considered within an organization, involves the development of a capability into a skill or ability. The operation of new equipment, technologies, work processes, etc. requires various types and degrees of skills that did not exist within the non-automated work environment. From the organizational point of view, it is necessary to train employees and managers in the skills necessary to operate in the automated work environment.

Training, however, is not just the development of new skills, knowledges or abilities. It should be part of the implementation process itself. Training curricula should aim to develop the employee/managerial skills and attitudes appropriate to the new work environment, as well as build technical skills. Unfortunately, this approach to training does not occur with sufficient frequency. The emphasis, more often than not, is on strictly technical training.

1. Background

ACS managers and employees were strongly supported by extensive and varied training in and out of the classroom, including hands-on telephone and terminal exercises, simulations of taxpayer conversations and on-the-job coaching.

Training was recognized as a critical process for implementation of the Automated

Collection System, but the pilots showed that the resources necessary for training managers and employees were significantly underestimated. Corrective actions, additional training and ongoing revisions and refinement doubled and quadrupled the original estimates of resources needed for developing, conducting and participating in training.

There were several reasons for these inaccurate estimates. One was the service's lack of experience in fully integrating new microcomputer technology with human resource needs and technical work systems. There was also insufficient recognition of the fact that the new collection system would continue to evolve and change as much as it did, resulting in the need for new information briefings and training as the system procedures changed and the bugs were reduced.

One of the clearest lessons learned was that training for complex systems should be planned as an ongoing process rather than a one-time event. The training changed as the system was adjusted and readjusted. On-site visits and narrative summaries of the attitude surveys showed that managers and employees felt overwhelmed by the extent and frequency of these changes. Modifications necessitated additional classroom training, extended on-the-job training, and reinforcement and revisions to procedural instructions to managers and employees.

While the IRS training and development organization was involved in the planning and implementation of ACS, there was a need for earlier involvement in the initial design stages of the system. Experience with ACS and other computerization efforts demonstrated the need for building in human factors and organizational considerations into each stage of the system development life cycle for new technology.

Training officials did work together with hardware, software and outside training contractors to develop procedures to help employees acquire new knowledges and skills. However, the ACS project was clearly driven by the technology and the time schedule for installation of machinery. No amount of planning eliminates all the needs for adjustments in a new technological system, but the more changes and refinements required after planning and initially implementing a system, the more revisions have to be made to the training designed for the system. Collection officials conducted extensive planning but found that it was not sufficient to avoid substantial changes after implementation.

2. *The ACS training plan*

The training plan for ACS pilot districts included preclassroom orientation and training, classroom technical training, procedures and equipment training for call site employees, and management training on the new ACS system, procedures and equipment. More limited training was given to service center employees.

In total, seven weeks training was given to technicians and up to eight weeks of training was offered to managers at the pilot sites. After implementation additional training was made available. Responses to questionnaires, training evaluations and reports, and on-site visits showed that the range and extent of training made a significant contribution to preparing for the new and redesigned collection work.

Both line and staff officials recognized that training can be a major organizational strategy for translating concepts behind the productivity goals of new technology and changed work systems into actual work practices in the organization. This is in contrast to organizations concentrating only on the most obvious training objectives such as computer operations and job-specific procedures.

Training and orientation programs can also shape attitudes about the system, instill new organizational philosophies, and assist employees accept change. Both the content of training and how it is presented expose managers and employees (particularly new recruits) to the organizational culture and any new expectations of the redesigned work. Consequently, **training should not stop at technical knowledge and skills. Planning for training should incorporate the full scope of an organization's needs, goals, philosophy, cross-functional relationships, norms and values**.

The Human Resources Technology Staff built on suggestions originating from early ACS task forces and training and development division officials to provide managing change, stress management and problem-solving courses for managers and employees, with job redesign training for managers alone. Some of the training included a *Deskercise* program designed to help relieve the stress and strain that can result from long periods of work at computer terminals. The HRT Staff also designed an orientation process to supplement a video-tape the service had used to demonstrate the operation and potential benefits of the ACS to employees.

These orientation briefings, materials and guidelines expanded the information originally developed for the ACS by giving ACS employees and new recruits a definition of the mission and purpose of the new system, an explanation of how their jobs fitted into the broader picture with other offices, and the human resources implications of computer-mediated work. Some of the readings duplicated articles given to field and national office executives at a previous IRS conference in October 1982. The readings accompanied workshop discussions aimed at having executives identify the organizational change issues likely to be encountered in moving towards a new and different form of work environment in the next decade.

Kick-off speeches were written for local executives to open the scheduled manager and employee training. These messages were designed to help employees understand the potential benefits of the new system and to focus on their roles for identifying improvements and solutions to problems encountered in their work. The executive comments described the ACS as a comprehensive, flexible computerized system using modern telecommunications technology to collect delinquent taxes. The introduction to the training also contrasted the benefits of the ACS with the problem-ridden Collection Office Function which preceeded it. A letter signed by the IRS Commissioner was included in the ACS pilot sites' training materials to signify the importance of ACS as the flagship in the service's efforts to accelerate the expansion of automation.

While planning for the ACS, managers and employees alike suggested that there might be a high degree of computer-phobia that could interfere with successful implementation of the ACS. This was forecast as most likely for those older Collection employees who had lived with the manual batch system for years. These concerns led to the development of guidelines for the pilot sites from the HRT staff on how to expand computer literacy and familiarity with automation (such as hands-on demonstrations of personal computer operations, visits to automated offices, and selected readings introducing computer operations in general).

The objective of the activities recommended and implemented by the Human Resources Technology staff was to focus on the human factors issues involved in the ACS. These included efforts to help people manage the organizational and personal changes affecting them as they assumed new jobs and duties and operated microcomputers. The activities were also intended to encourage innovation and problem-solving by employees at all levels as they worked to increase the collection of delinquent taxes (see Figure 8–1).

HRT staff members also addressed the managers and employees at the beginning of

Figure 8-1. Objectives for the courses developed by the HRT staff in cooperation with training.

Managing change for managers

At the end of this lesson the participants will be able to explain how to diagnose and deal with typical feelings and behaviors associated with stages of the change process. This will be accomplished by:

1. Identifying, discussing, and analyzing issues, problems and concerns which they feel are critical to the success of the ACS.
2. Identifying ways of managing the work force during a change.
3. Identifying and discussing issues associated with implementing change.

Managing change for employees

At the end of this lesson participants will be able to describe:

1. Some typical feelings and behaviors associated with change.
2. Some techniques for coping with change.

Stress Management for managers and employees was designed to help them identify signs of stress and learn psychological and physical coping techniques to reduce the negative impact of stress.

Problem identification and problem-solving techniques were woven into managers' and employees' training to help them deal with ACS issues.

training classes to explain the work of the group, to reinforce the fact that the IRS was concerned about the people issues, and to administer the first attitude survey in their research plan.

3. *Training functions and objectives*

Another clear lesson from the ACS implementation is that **training for new information systems could serve broader and more basic functions beyond technical training**. Examples of expanded functions are the following:

Specify the objectives and advantages of a proposed system.

Help shape attitudes about the system and gain acceptance of organizational change.

Communicate managerial goals and philosophies.

Identify shifts in roles, activities and relationships between offices, including the redesign of organizational structures.

Communicate ongoing changes to the system.

Communicate actions designed by management to make the transition less difficult (such as staffing and redeployment flexibilities, specific skills training, cross-functional technical training, and opportunities for identifying and resolving problems).

This kind of training and orientation can increase receptivity to change, answer the typical concerns inherent in any change, and bring to the surface resistance, questions and suggestions that may have been overlooked by the planners, or that still require the attention of the project team.

It can also help managers and employees learn what will happen in the transition stage, as well as what the redesigned work system will look like when it is implemented. In order to obtain the maximum benefit from this approach, orientation and training should be a two-way communication process so that ideas from workers flow back to management and concerns are dealt with in a timely way.

The need for and value of this broader approach to orientation and training tends to be most evident when it is absent and employees voice their feelings of isolation and lack of knowledge about the work of the organization. Some officials questioned the need for information beyond the immediate job duties of the employees' assigned work. Employees tend to have a fragmented view of the work and are left unprepared by management to improve work flow, release bottlenecks and maintain high quality if they do not know what happens to the work before and after it crosses their desk.

Team building as an adjunct to training

The pilot sites and subsequent ACS sites used IRS specialists (in-house consultants) or contracted external consultants to assist them in conducting team-building sessions for their managers. These were designed to acknowledge the demise of the Collection Office Function and prepare for the ACS as well as to bond members of the new work teams together. Managers used these meetings to clarify local goals, to establish and negotiate roles and support for one another, to solve problems, and to develop actions plans to implement ACS effectively and efficiently. (Team building is covered in greater depth in Chapter 9.)

ACS training findings from the third questionnaire

Training was evaluated by classroom reaction sheets from participants, instructors and on-site observations by Collection and Training staff members. In addition, one pilot site implemented an evaluation of training and on-the-job performance a few months after the initial training to assess how well employees and managers were able to accomplish their assigned tasks without assistance. The findings in this section resulted from the third questionnaire.

Employees at the ACS pilot sites reported significantly higher satisfaction with their training than the comparable offices at Collection Office Function sites. However, less than 80% of the respondents were fully satisfied with the training and on-the-job instruction in the sites.

Both employees and managers indicated that they needed a greater degree of technical training (e.g., tax law and collection procedures). There were many reasons suggested for needing more training. One was the different backgrounds of employees and managers in terms of their relative familiarity with the technical tax field. Another was that the tasks to be performed by managers and employees were not clear during the planning stage. There were ongoing changes during and after implementation to help clarify these tasks and support them with training. Managers also reported different interpretations of the new collection philosophy (emphasizing taxes collected rather than cases closed) and how philosophy could be translated into concrete instructions and techniques for effective tax collection.

Satisfaction levels for *classroom training* varied considerably from site to site, ranging from 81% satisfied to 49%. Sites reported the need for reinforcement to strengthen training. The extensive length and relatively complex training required too great a demand for both acquiring technical knowledge and building new skills in working on automated equipment. Employees new to the IRS reported greater difficulties in trying to learn about the IRS, the Collection Function, and how to use the new system.

ACS employees rated the non-IRS training provided by the contractor or a vendor as less satisfactory than the IRS training. Although there was a high rate of non-response to this question, the vendor-provided training was rated as satisfactory by only 61–69% of the employees at the four call sites.

On-the-job instruction satisfaction levels ranged from 56% to 89% at the sites. The sites indicated a need for improvement in training manuals and references (with dissatisfaction levels ranging from 24 to 30%). These responses may reflect some of the same issues uncovered in the classroom training where new materials had to be updated frequently as the system was refined.

Employees had mixed reactions to the amount of training in five different categories, with some employees rating the same categories as "just right", "too little", "too much" or "not relevant". All sites indicated they wanted more tax law training.

Feedback on the usefulness of managing stress and managing change training on the job was mixed. The training covered techniques for application both on and off the job. There were high degrees of satisfaction reported on evaluations completed in the classroom and follow-up telephone surveys, with strong recommendations to continue the training. However, respondents to the third questionnaire reported that they had little or no time on the job to practice the coping techniques they had learned. They recommended that refreshers and reinforcement be provided to ensure practice.

There were several factors which may account for the variations among the sites in their reported satisfaction levels with training. These would include the following:

Differences in local training administration practices.
Variations in the effectiveness of instructors and coaches.
Degree of managerial involvement in training.
Differences in organizational climate and training settings.
Differences in pre-existing attitudes among trainees.

Suggestions for modifying training and orientation

Substantial attention was devoted to identifying and resolving training-related problems when they occurred during the pilot phase of the ACS. Significant changes were made to the pilot training materials for sites scheduled for implementation after the pilots. Modifications were made to refine training approaches, help solve identified performance problems and incorporate new and revised procedures in technical and managerial training. Managers wanted more coverage of problem-solving techniques, how to counsel employees, administrative procedures for the new system and the Rockwell telephone system.

Organization and management factors

There were several factors related to management decisions which affected the training plan. For example, a significant number of procedural and management report details were unresolved during the time training was developed and delivered. Trade-offs were made because the service emphasized bringing the technical system on-line on schedule, and numerous modifications to training were made as procedures and job responsibilities were clarified.

The field offices indicated employees' and managers' concerns about the lack of contact between supervisors and their employees prior to going on-line. This occurred in part due to the scheduling plan rotating around the installation of equipment, and the need for some managers to acquire more knowledge in collection techniques, tax law and the new system. Employees tended to have much more contact with their on-the-job instructors and little contact with their supervisors. This led to some role ambiguity, with employees viewing the instructors as helpers and being somewhat unclear about the roles of their managers.

This situation can easily be avoided by considering different schedules so that managers are trained before employees and freed to lend support in training, or by using a team approach when the training material for managers is similar to that for employees.

Recommendations to orient and prepare future call site personnel

ACS field employees and managers suggested the following actions to help prepare future ACS call site personnel:

Sponsor orientation visits to a working ACS call site.
Provide periodic briefings on the implementation of the ACS.
Arrange for visits to other automated offices and readings in computer literacy.

Recommendations for future training actions

Project managers were not prepared by either experience or training to assume major project management tasks. Consequently the Human Resource Technology Staff has recommended project management training to help future efforts in planning and implementing automated systems.

Managers would also benefit from training and organizational reinforcement in the following areas:

Change management: leadership and management styles.
Organizational effectiveness: relating social and technical aspects of introducing automation.
Advanced techniques for improving motivation and productivity.
Expanded skill-building of computer competencies.

Feedback from on-site visits to the field indicated that changes in philosophy and major changes in the ways of doing business need continuing monitoring and reinforcement. For example, ACS managers wanted to ensure that employees were 'firm but fair' in dealing with taxpayers, and sufficiently aggressive to ensure that delinquent tax payments would be paid immediately. Although a large number of simulated cases were developed for telephone techniques training, managers wanted to make sure they would represent realistic assessments of difficult cases.

There were ongoing projects to identify effective telephone collection techniques and a greater number of role plays for continuing practice of employees. Self-paced programs were incorporated into the initial pilot training. The HRT Staff recommended that self-paced instruction be expanded to link what had been in classroom training, information briefings and meetings and role plays.

In addition to having on-the-job coaches and instructor manuals for these employees, the HRT Staff recommended the establishment of an 'experienced employee' buddy system. This would entail each new employee being paired with an experienced employee who would be responsible for the development of the new employee as part of his or her performance evaluation. (This is standard practice in some functions that are well-established and is institutionalized by including these duties in position descriptions for senior employees.)

In view of the limited task variety and lack of a perspective of the total Collection Function, the group recommended cross-training for employees on rotational details through all Collection Functions and to the service centers servicing call-site districts.

HRT suggested that a continuing professional education program be established for ACS managers to incorporate topics such as performance appraisal and effective feedback, motivational techniques, problem solving, group dynamics and leadership techniques.

Other suggestions for training were as follows:

Monthly seminars on selected human resource topics.

Case studies with exams for new employees that are scored by their managers to help the supervisors assess the competence of employees.

Videotapes of employees successfully operating workstation furniture, equipment and proper techniques.

Production of a monthly newsletter to heighten awareness of human resource issues and to provide practical guidelines for managers (examples of topics include: setting goals and objectives; counseling employees; motivating; promoting creativity; giving and getting feedback, etc.).

Other recommendations based on lessons learned from ACS training include the following:

Clearly define managers' expectations and employees' tasks prior to training to avoid excessive readjustments and retraining after implementation.

Address the relative costs and benefits of refining the system and work procedures before and after implementation during the system's planning phase; and ensure that the implications of the decisions on which course to follow are understood by decision makers.

Build in time for supervisors to get involved with their employees during the orientation and training process to convey managerial expectations and build effective work teams.

Build in sufficient flexibility in training plans to meet the needs of individual users at different geographic locations.

Build in adequate resources for training and for guiding and working with training contractors into system development plans (working with contractors may double the initial time allotted for vendor-provided training in view of the need to orient them to unique organizational issues and procedures and keep up to date with system design changes).

Develop additional training for managers based on their role in the new information system such as project management techniques; socio-technical analysis of work systems for redesign of jobs; effective management of change; and improving motivation and productivity.

Chapter 9
Team Building and Resources

Introduction

As is the situation in many non-automated environments, the work team forms the core element determining unit productivity, job satisfaction and group morale. While efforts aimed at improving the effectiveness of individual performance can result in some impressive gains, the effort involved in such activities is high and the success rates relatively low. In contrast, the pervasive effects of the work group on individual behavior are well-documented. In terms of time and effort expended on improving organization productivity and job satisfaction, a focus on the work team has consistently resulted in positive outcomes.

Depending on its function, automation can serve to either increase or decrease the interdependence of group members. In the case of increased interdependence, the likelihood of destructive conflict and competition among group members tends to increase. In the case of decreased interdependence, a sense of interpersonal isolation and lack of group goals is commonplace.

If one of the effects of automation is to reduce the potential for intrinsic job satisfaction from the work itself, the influence of the work team on individual job satisfaction and productivity increases. Where the tasks are short-cycle, repetitious, and require little concerted intellect to perform, the employee will often turn to his or her co-workers for the support, reinforcement and stimulation lacking in the work itself.

In either situation, the need for an effective work team is paramount. The common reaction of management regarding the effects of automation on the work team is to increase the degree of control. This reaction frequently has a boomerang effect by strengthening co-worker relations in uniting them against a common 'enemy', the boss. Since the supervisor rarely, if ever, performs the tasks assigned to the rank and file employee, there is little understanding of its effects on the individual and work climate. Employees' reaction to this are, predictably, negative.

1. Background to the Focused Group Interview

A variety of related approaches to improving work group effectiveness are available. For the most part, the available programs appear to focus on one aspect of the team (e.g., communication skills, group conflict, problem solving, etc.). While few group processes are independent of each other, the focus on a single set of issues was felt to be less comprehensive than desirable. This was especially true given the relatively limited access to work groups that was available. The cost to the organization of getting a group of employees together for more than a brief discussion was considered high.

Given the number of group aspects that needed to be addressed in a brief period, a program based on a highly modified Force Field Analysis approach (Lewin, 1947) was developed by the Human Resources Technology group. The opportunity to apply our "Focused Group Interview" (FGI) approach arose several times over the course of ACS implementation. In each case, the goals were slightly different but the procedure proved sufficiently flexible to allow for modification to meet the needs of the specific situation at hand.

In practice, a focused group interview can be used to accomplish several different objectives within the same frame. Another advantage is that it is very much centered on the accomplishment of work. "Touchy-feely" clinical approaches to team effectiveness are, generally, not as well accepted as are efforts directly aimed at building productivity. A third advantage is that the technique does not require a great deal of psychological insight on the part of the people administering it or content knowledge of their function. In fact, a dearth of knowledge concerning the function can frequently be valuable since it requires the facilitator to question employees and have them explain assumptions that otherwise may go unquestioned or assumed.

The ability to draw people out, provide an open environment for discussion, and get individuals to express themselves in a practical and concrete fashion is, however, very valuable. These skills, depending on the individuals involved, are not difficult to master with a few sessions of observation and practice.

Lastly, the FGI technique had an advantage over other methods because it proved to be relatively inexpensive to conduct. Since it is based on a logical sequence of activities, it can be easily learned and administered. The time required to administer it varies from one to three hours depending on the goals and experience of the facilitators. It requires no equipment other than a pad of paper and pencils, a simple form, and a blackboard or flip chart.

In its broadest use, the Focused Group Interview can be used to accomplish five independent goals within the same context. Depending on the amount of time available and interest expressed, the actual use of the technique may involve as few as one objective and partial accomplishment of others. The five objectives that may be achieved by using the technique are:

1. *Problem identification*. Identify major strengths and weaknesses of current or planned activities, policies and procedures.
2. *Problem resolution*. Identify alternative problem solutions and develop action plans to implement those solutions.
3. *Data collection*. Collect quantitative and qualitative information with respect to employee and managerial attitudes.
4. *Model effective group behavior*. Provide guidance and feedback on effective and ineffective ways to interact and solve problems.
5. *Build team morale*. Generate a participative and positive climate among team members.

An indirect side benefit of a successful FGI is its effect on the management style of the organization. A commonplace fear among managers and executives is that employee participation and input into decision making will, inevitably, result in organizational anarchy or a 'country club' atmosphere. The results of the FGI approach have frequently provided evidence for the opposite conclusion; that employee involvement in problem identification and problem solving is an effective method for improving productivity and morale. When faced with the issue of how to make their work 'better', employees almost always focused on improving the efficiency of the work flow, improving feedback on performance, reducing conflict and getting more done. The practicality and task orientation of ideas generated from employees have, in several cases, altered the attitudes of managers and executives towards employees.

The Focused Group Interview is thus a useful and time-saving method for collecting information regarding employee attitudes on a wide range of topics. Although it can be used to assess individual attitudes, we have found it more appropriate when conducted in a small group setting. In comparison with other information collecting techniques such as interviews and paper-and-pencil surveys, an FGI combines some of the advantages of both. An FGI provides an approximation of the quantitative information available through closed-end questionnaires, as well as some of the qualitative, in-depth data that can be gleaned from one-on-one, unstructured interviews.

Useful as this technique is, it should be used in conjunction with, and not as a substitute for, other modes of data collection. It can provide a valuable first step towards identifying issues for follow-up research or as a procedure for elaborating/confirming issues identified with other types of research methodology. Whatever its limitations, the economies of time and scale and its orientation towards employee involvement in problem-solving with the FGI compensate for its limitations.

2. *Focused Group Interview procedures*

To conduct a focused group interview, participants are gathered together in a single room for a period of 2–3 hours. The amount of time required will vary with the number of participants and, most importantly, the number of study objectives as described previously. No more than 25–30 participants should be included in one session. If possible, participants in each session should be drawn from the same work group, functional area, or common interest (e.g., all managers). Participants should have had sufficient prior notice of the session and its goals to schedule it within their work schedule. If several sessions with different groups are planned, it is important that the first session conducted go smoothly since the grapevine 'word-of-mouth' communication among employees will tend to exaggerate the success or failure of the session. We have achieved better results (i.e., less inhibited discussions) when peer groups of co-workers and supervisors meet separately during the first meetings. If appropriate, supervisors and their employees should join the session at a later time.

At the beginning of the session facilitators should introduce themselves and the objectives of the meeting. The anonymity of the participants' comments during the session should be assured.

Two facilitators should be used for each session. Although the roles of the facilitators will change during the session, one generally acts as a recorder and the other as a discussion leader.

The session should proceed within the following guidelines:

Step 1. Individual ratings and factors (10–20 minutes)

A. To set the focus for the session, an FGI response form, illustrated in Figure 9–1, is handed out to participants.

A broad, generic issue is identified along the base-line of this form. For example, if the study is designed to investigate staff turnover, the broad issue identified along the base-line might be "Likelihood to Remain in the________Function". For this issue, the scale might range from "0" equal to "No chance at all", through "4" equal to a "50/50" chance to "7" equal to "100% certain to stay".

B. If appropriate, individuals are asked to indicate on the FGI form some information which might be relevant to the study. For example, information regarding age, sex, length of service, grade or job series might be collected. In most cases, it is inappropriate to require individuals to write their name on the form although they might be identified indirectly through the other information requested. Again, participants should be assured of the confidentiality of individual responses and that results will only be reported in a global fashion.

C. After providing the appropriate background information, participants are asked to consider the issue identified on the FGI form and, to the best of their abilities and current state of knowledge, make a judgment regarding the answer most representative of their own assessment. They are to circle that number. No more than five minutes should be allowed for this.

D. Once the participants have completed Step C, they are asked to indicate the

Figure 9–1. The Focused Group Interview form.

Background Data:

Age: ______
Job Title: ______
Grade: ______
Years in Current Job: ______

PROMOTING FACTORS	*RESTRAINING FACTORS*

1	2	3	4	5	6	7
extremely low			average			extremely high

ISSUE STATEMENT: ______________________________

Figure 9–2. An example/demonstration FGI form.

PROMOTING FACTORS	*RESTRAINING FACTORS*
Feels good when I exercise	Takes a lot of time
Meet people when jogging	Detracts from other chores
Improves mental attitude & outlook	Depends on weather conditions
Come back to work refreshed	Neighbors laugh at me (embarrassed)
Improves health/weight loss	Sometimes too tired
	Bike is in bad shape

1	2	3	4	5	6	7
			×			
extremely low			average			extremely high

ISSUE STATEMENT: Current level of Physical Fitness

factors which figured in their assessment of the issue. If they rated the issue a 5 on the scale, what factors kept it from being a 1. These 'promoting' or positive factors should be listed on the left-hand part of the form. If the issue is turnover, for example, those factors which have an effect on their remaining in their job should be listed. There is no limit to the number of factors that can be listed by the individual although, as a matter of practical concern, people tend to list between three and ten factors.

E. On the right-hand side of the form, participants are then asked to list all the negative or 'restraining' factors which act to restrict or lower the maximum level of the issue. Put another way, "What keeps them or the organization from being a 7?" The same factor may be listed on both the positive and the negative side. For example, 'training' might provide the necessary skills to accomplish the job effectively (i.e., a positive factor) but, in addition, provide individuals with marketable skills which increase the likelihood of them leaving the organization (i.e., a negative factor).

A prepared example often helps the participants understand the process of identifying promoting and restraining factors. An example that has been effective in demonstrating the process revolves around the issue of degree of physical fitness. The completed example, illustrated in Figure 9–2, is exhibited before participants while the facilitator 'walks' the employees through the concept of assigning a rating to the degree of physical fitness scale and then identifying the factors which promote the current level and detract from optimum physical fitness.

Step 2. Small group consensus (30–40 minutes)

A. After the participants have individually completed Step 1E, they are asked to join together in groups of 3–6 people. The task of each of these small groups is to complete a possible list of all the promoting and restraining forces they have identified individually. All duplicated items should be listed only once. New items may be listed which don't appear on the individual sheets but are the result of group discussion. The facilitator should hand out a blank FGI form to each group to compile the group's master list. If the small groups comprise members from the same functions or work groups, the group's designation should be listed at the top of the group FGI form. It is helpful, although not essential, that these small groups are from the same team or function.

B. Employees are instructed to share their individual lists with the small group. One person in each of these groups should record all responses on the master list, eliminating duplication.

Step 3. Large group consensus (40–60 minutes)

A. After all groups have completed their composite list, a representative from each group is asked to read aloud the group's list of promoting and restraining factors. At the option of the facilitator, the length of the session may be shortened by emphasizing only the restraining factors. As each group reads aloud its items, the other groups are instructed to review their own lists to eliminate duplicate items. As the group reports progress, typically fewer and fewer new items are reported.

B. The facilitator's role during this period is to act as a naive sounding board for the groups' presentations. He or she does not challenge, filter or voice any judgment with respect to the items. The facilitator asks questions to clarify and amplify the responses so that everyone has a common understanding of the forces being cited. This is done by questioning (e.g., "What I think you're saying is . . . ", "Could you give me an example of that?", "How is what you're saying different from the factor the other group mentioned?", etc.).

In clarifying and elaborating the factors identified by each group, a key role of the facilitator is to force the participants to make abstract responses more concrete. For example, the response 'Poor management attitude' should be made more explicit by having the specific behaviors representing poor management attitude.

C. As one facilitator focuses on clarifying the group presentations, the other facilitator keeps a running list of the items reported on the blackboard or flip chart. The facilitator should also attempt to identify similarities among factors, forming categories of item clusters when these occur.

Although the number of positive and negative factors identified will depend on the issue being discussed and the emotionality of the groups, typically 15–25 restraining factors and 10–15 promoting factors are cited. There is, of course, no correct or incorrect number of factors. As the groups present their items in a round-table fashion, new items, not identified on any group list, may also be added to the master list developed by the facilitator.

D. Following the last of the group presentations, the facilitator remarks that some of the items on the master list appear to be more important than other items. To reduce the number of items to a more manageable number, the groups are to reform and discuss among themselves which items from the master list are more critical to them. Of all the items detailed on the master list, they are to choose the three most important. Allow about ten minutes for group discussion and voting. It is often helpful to facilitate this prioritizing process offering two criteria for selecting items:

1. Which three factors have the greatest impact on the issue being examined?
2. Of those which have a significant impact, which are also susceptible to solution at the local (or whatever appropriate) level?

Because of limitations of time, attention and interest levels, the facilitator may want to limit the discussion and ranking at this point to the 'restraining' factors. As a general rule, participants tend to be more motivated to deal with the negative rather than the positive aspects of the work issue.

E. Once the groups have reached consensus, each group is asked to report the three factors they have identified as most important to the issue being examined. During this step in the process, duplicate responses are acceptable. Each time a factor is cited by a group, a check mark is made next to that factor. It is a common occurrence to find wide agreement among employee groups on which factors they identify as being the most important.

The session may be ended at this point by thanking the participants for their time and effort. The FGI sheets filled out by the groups and individuals should be collected at this point.

One of the most frequently asked questions concerns feedback on the study's findings and recommendations. Whether or not employee and/or managers will be given feedback at a later time is an issue that needs to be dealt with prior to the data collection. This decision should be based on the study's goals, the people involved, and the effects that feedback (or its lack) will have.

Step 4. Group problem solving (30–90 minutes)

A. After the major problems have been identified, groups are asked to develop plans to improve the situation. Each group is assigned a different problem which has been rated as 'important' from the master list developed in Step 3E above. The criteria for these plans are:

1. The plan must be practical in the sense that it doesn't involve a major disruption of the workforce, workflow, personnel changes or costs.
2. The plan must be practical in the sense that changes, if contemplated, must be within the authority of local responsible officials. For example, pay scales are not determined locally.
3. The plan must be short-term and focus on steps to be taken within the next week or two.
4. The plan must specify who is to be involved, what their specific responsibilities within the plan are, the time-frames involved, the resources needed for each step, and what, in a concrete sense, is to be achieved at each step of the plan.
5. Where possible, back-up or contingency plans are to be made at critical stages of plan implementation.

B. Groups are reassembled. Each group is assigned (or chooses) one of the most 'important' issues identified in Step 3E. The groups are then given 20–30 minutes to develop an action plan that adheres to the criteria set forth in Step 4A above.

The time frames allowed for Steps 4A and 4B are purposely kept short and should be adhered to strictly. Allowing extensions of time or longer time periods appears to foster 'nit-picking' and attempts to define too much detail during this phase of the FGI. By keeping the time frame at a minimum, participants are forced to focus on the broad outlines of the task.

C. At the end of the allotted time, each group, in turn, presents its plan to the other groups. If time allows, it is helpful to have each group write down a draft outline of the major components of its plan and have these distributed to the other participants.

During each group's presentation, all participants are encouraged to question or comment on the plan being presented. Discussion should, however, be kept to a minimum by keeping the focus on the broad outlines of the plan with the comment that "details can be worked out later". Applauding after each presentation provides a great deal of motivation for the employees at this point.

D. Following the presentation of all groups, the facilitator should lead a discussion regarding the possible conflicts, redundancies and overlaps between plans. The goal here is to get the groups to share resources, provide back-up to each other, and work together to achieve their goals.

E. Each group is asked to formalize its proposal in writing. These proposals should take into account the comments made by the other participants during the presentations. Copies of these proposals are to be made available to all group members.

F. It is extremely desirable to have the group members present their plans before a top-level executive committee for review and approval. Since the initial enthusiasm tends to dissipate rather quickly as people get drawn back into their day-to-day routines and emergencies, it is desirable to minimize the time between idea, discussion, and presentation—no more than a week or two from initial meeting to presentation.

Presenting their ideas and plans before top management not only has the effect of promoting a climate for change but, additionally, assuages the concern of some managers regarding the abilities, practicality and productivity orientation of the majority of employees.

G. If the plans are approved for implementation, the group that devised the approach becomes the task group charged with its implementation.

The facilitator should emphasize that no plans or expectations will be implemented perfectly. It is a common reaction among group members to give up if one aspect of the plan doesn't work out as they had planned. The purpose of back-up plans for critical aspects is the anticipation that changes will occur.

Step 5. Follow-up support

However much enthusiasm is generated during the initial steps of the FGI process, the motivation to maintain efforts necessary to implement the elements of the plan decreases over time and obstacles.

It is highly advantageous to schedule a follow-up session with the participants. The timing of this follow-up session is important. It should be somewhere around the middle of the timeframe for implementation since this is the point at which frustration is generally highest. The session should consist of a review of project plan accomplishment and obstacles. The role of the facilitator should be one of active support by encouraging the participants to think through alternatives available or changes in strategy. At times, the facilitator might act as a go-between among managers, group members and task groups, although it is desirable to force a confrontation between conflicting parties. In these cases, the facilitator should act as a neutral 'translator' by trying to get different parties to understand the position of others.

3. Applying the Focused Group Interview technique

The Focused Group Interview technique proved useful in a variety of situations. Its usefulness was a function of three aspects:

1. It was focused on the work, not on personalities and was thus much more acceptable to managers and employees.
2. It was flexible enough to be used in a variety of situations, changing it as the circumstances warranted.
3. It did not presuppose a specific set of questions and answers which might bias the study results. Rather, it started with a very broad issue and let the participants develop the questions and responses.

Although the Focused Group Interview technique was used in a number of sites to explore a variety of issues, a single example will be presented here. This example is typical of others in most respects, but it should be borne in mind that each application of the technique yielded some results unique to that place and time. In describing the FGI results, we have attempted to be as general as possible.

In January, 1985 a request was made by a District Director for the HRT Staff to investigate the 'burnout' among the management staff at an ACS call site. To check this claim two members of the HRT Staff conducted several FGI sessions with both employees and managers.

Our activities with both managers and employees had a common rationale and direction: to help those responsible for the success of the call site to move from vague uncertainty and a sense of powerlessness to clarity about the basic issues and solutions and a sense of empowerment to implement those solutions.

Methodology

The Focused Group Interview technique was used with three separate groups of employees (i.e., one for each function: Contact, Research and Investigation) and one group of mid-level managers. The ACS Branch Chief was not included in the FGI sessions but was present during the presentation of ideas. With both the employee groups and management group, we focused on identifying the fundamental problems and issues that needed to be addressed; prioritizing the problems and issues; and generating realistic solutions.

The timing and sequence of delivery of the activities which represented these three elements differed for the groups of employees and managers. The FGI sessions were accomplished in the following order:

1. Meet with ACS Branch Chief and Collection Division Branch Chief to brief them on the processes we were going to use.
2. Interview ACS mid-level managers as a single group.
3. Interview ACS employees in groups of functional specialty.
4. Meet with ACS mid-level managers again.

For the employee groups, the problem solution phase occurred immediately after the prioritizing exercise. With the managers, by contrast, we delayed the problem solution step until our second session with them. In addition, the managerial group was provided with feedback about the results of the employee group discussions.

For the beginning problem identification phase of our sessions with both employees and managers, we used the FGI response form illustrated in Figure 9–1. As discussed in the description of the FGI steps, we asked for employees to indicate an overall rating of the effectiveness with which the work was done at that call site and for a list of factors that promote and restrain the work process. With all groups we conducted this problem identification exercise in three cycles:

1. Individually.
2. In small groups where individual assessments were combined, expanded, and collapsed.
3. With the entire group to establish a composite list of each group's responses and for ranking the most important items.
4. (For managers only) Develop an action plan for reducing the key restraining forces.

Employee findings

As a general conclusion, the employees with whom we spoke appeared to us to be motivated and professional. They had a good understanding of how the ACS operated and its potentials. With few exceptions, they expressed the strong desire to 'get the job done right'. Whatever problems existed in the call site were not likely to have resulted from low employee motivation.

In all, about 40 restraining factors were cited by the different employees groups. These factors fall into three broad areas:

1. Factors dealing with the interaction with and treatment by managers.
2. Factors dealing with bottlenecks or inefficiencies in the way the work gets done.
3. Factors dealing with the physical circumstances of their work.

The groups of employees were asked to prioritize these 40 odd factors. The listing, in order of ranking is given in Table 9–1.

Table 9–1. Listing of restraining factors identified by different employee groups, in order of decreasing importance.

1. Management style
   - An emphasis on quantitative standards to the detriment of quality issues.
   - Treating employees like infants.
   - Excessive negative feedback and corresponding lack of positive feedback and recognition.
   - Little opportunity for employee input and a sense that suggestions 'go nowhere' or, at least, are never recognized or reported back on.
2. The volume, pace, and timing of changes in policies and procedures.
3. Poor systems for communicating and disseminating important work related information. This reaction was compounded by factor 2 above.
4. Inconsistencies in management interpretation of policies and procedures resulting in inconsistent approval standards, inefficient approval processing, 'shopping around' by employees until they found the answer they wanted, and a climate of excessive caution and anxiety for employees.
5. Restrictions in pay, grade change, and career ladders for career growth.
6. Excessively structured work processes and routines resulting in a heightened sense of isolation, lack of teamwork, devaluation of employee skills, and no input into the decision making process.
7. Physical security of the site, especially for non-prime time employees.

The order in which the factors were ranked strongly indicates that employees were concerned with their treatment by managers and factors which inhibited work effectiveness. Several employees commented that they were limited in their ability to 'get the job done' because the new computerized management systems emphasized quantifiable results. Fruitless activities were being encouraged simply because they were part of the measurement/evaluation process rather than contributing to the work flow. The facts of the matter aside, many employees viewed this situation as the classic misapplication of new technology.

Managerial findings

The procedure for identifying and ranking factors was identical in the managerial and employee groups. After conducting sessions with both groups, it was clear that a consensus of opinion with respect to the restraining factors existed among all groups of employees. In general, managers showed a great deal of concern regarding the needs of their employees. Given the employee concerns, the sensitivity of managers to employee needs apparently is not being communicated downwards.

Managers as a group identified 37 restraining factors. These factors fall into four categories, listed in Table 9–2.

Table 9–2. Categories of restraining factors identified by managers.

1. Those that deal with the interaction among managers.
2. Those that deal with employees' relationships with their managers.
3. Those that deal with ineffective work processes.
4. Those that deal with the physical work environment.

In the same way as the employee groups, the managers were asked to prioritize the 37 listed factors, listed in order of their importance in Table 9–3.

Following the development of this prioritized list, managers were encouraged to develop short-term plans for the amelioration of the top five problems. Within a very brief time-span (20 minutes) they were able to generate a general outline of an

Table 9-3. Listing of restraining factors identified by different management groups, in order of decreasing importance.

1. Poor long-term planning resulting in continually being put in the position of 'putting out fires', 'minimizing disruptions' and going from 'crisis to crisis'.
2. The volume, pace and timing of changes in policies and procedures.
3. Poor systems for communicating and disseminating important and time-sensitive work-related information.
4. Lack of teamwork and co-ordination among managers.
5. Problems with the Service Center's timeliness of information entry and accuracy of that information.
6. Unreasonable production goals.
7. Lack of direction from upper management.
8. Grade and pay inequities among different functions in the ACS.
9. Lack of an appropriate and timely means for rewarding and recognizing the efforts of employees.
10. Lack of authority to delegate any but the most trivial administrative tasks.

approach. Each group presented their plans to the other groups for review and critique. The enthusiasm that was generated in this session went far beyond what we had expected.

Unfortunately, the presentation of the plans by managerial groups was not received with as much support as we would have anticipated. So far as we are able to determine, no substantive activities have been implemented as a direct consequence of these sessions with managers and employees.

With hindsight, the full commitment of executives and upper management levels needed to exist for any of the plans, no matter how imaginative or practical, to be implemented. The creation of expectations for improvement which occurred as a result of our sessions may have, ultimately, resulted in a higher level of frustration than existed prior to our arrival. The sessions, while cathartic, may have also tended to increase the intensity and mutual awareness of the pervasive issues that existed in the call site. In addition, it is our feeling that the expression of opinions within the context of a group of peers did much to create a feeling of solidarity toward common problems.

The initial issue which provided the impetus for our visit, 'managerial burnout', proved to be of minimal concern to employees and managers. While, in several cases, workloads were excessive and many duties were assigned which were collateral to the primary work requirements, this was a temporary condition due to the site start-up requirements. The deeper and more pervasive issues were those of values and attitudes which were reflected in practices and procedures. Unfortunately, to varying degrees, these values tended to permeate many of the call sites, possibly due to the very nature of the tasks people were required to perform. The ACS operation was founded on the necessity of collecting overdue taxes. This 'deadbeat' philosophy was expressed in a lack of trust, emphasis on accountability and measurement for control. This was never verbalized but, in the opinion of the author, represented a widespread organizational culture.

Chapter 10
Organization and Management

Chapter 5 briefly described the administrative structure of the ACS operation. On paper, each ACS call site was to be a clone of every other site. The intention was to standardize procedures and activities to facilitate compatibility. For the most part, this was accomplished. Each site operated under the same set of rules, regulations and position descriptions. This is not to say that there wasn't a great deal of variance among sites. These differences were due to ambiguities in the system which required on-the-spot decisions, a lack of timely communication among call sites and the National Office, and the desire for local sites to localize procedures in accordance with their unique situation.

During the initial stages of implementation, call site differences were either unrecognized or tolerated due to the immediate need to get the entire system up-and-running. Some sites, in fact, chose to ignore some National Office directives when they were in conflict with the local situation. Two of the widespread variations concerned the 60/40 rule and job assignment.

The 60/40 rule specified the employment ratio of full-time permanent employees to 'other than full-time permanent' (OTF) employees (i.e., permanent part-time, temporary or seasonal employees). The assumption underlying this ratio was that workloads would vary over the course of the year. By having 40% of the staff on a non-permanent full-time basis, staffing adjustments could easily be made to meet workload needs.

In reality, the workloads did not remain stable; the number of cases in inventory increased continually. Adding to this increasing backlog was the unexpectedly high turnover rate among ACS employees and managers. Many of those employees who were hired on an 'other than full-time' basis did not receive many of the benefits available to the permanent staff such as retirement, sick and annual leave. In accepting employment with the ACS, many of these employees expected to eventually become full-time permanent staff. In other cases, many of the OTF employees took the position as a temporary measure until a better position became available in another organization.

The workload levels and turnover rates in many of the OTF positions created problems for training, recruitment and staffing. In response, several sites created more full-time permanent positions. Such a response created a more stable workforce since many of the part-time employees were, in reality, working a full-time schedule.

The need to balance the workload demands and the availability of personnel resulted in some additional 'local initiatives'. One of the activities in which this workload–staffing level imbalance occurred was in the area of incoming and outgoing calls. Outgoing calls were made for the purpose of contacting taxpayers or third-party individuals (banks, employers, other agencies, etc.). Incoming calls were generally from the taxpayer or their representatives who were returning a call from the ACS. The number of incoming calls that were received far exceeded what was expected. It was hoped that outgoing calls would resolve the tax liability issue with little need for follow-up. The philosophy of 'one call does it all' was used to justify this belief.

Disparate staffing created by this unanticipated demand created some conflict within the sites. Since, in agreement with the union, position descriptions specified the exact duties of each job, it was technically not possible to require employees in one function to perform duties assigned to another function. In some extreme situations, employees in one function would be overwhelmed by incoming calls while those in another function would be working at their usual pace. Several informal local arrangements were made to rectify this imbalance. On a national level, a 'universal position description' is currently being tested which allows employees to perform all three functions (Contact, Research and Investigation).

Another administrative issue that arose during the initial stages of implementation was site size—the number of employees per site. The contractors, who helped design and install the system, recommended that sites should not exceed 100 employees. They believed that larger sites tend to lose the sense of community that many of their smaller sites have developed. The Collection Division established a task force which investigated this issue and its effects on the desired span of control within the call sites. Most call sites, as a result, average around 100 employees with a typical span-of-control of between 1 and 10, although there are differences among sites due to the nature of the geographical area they serve. Larger sites tend to cover a larger population base than do smaller sites.

Two other, somewhat technical issues of organization need to be cited here. The first concerns the Integrated Data Retrieval System (IDRS) and the second involves the process of case assignment.

As discussed in Chapter 5, the Integrated Data Retrieval System was designed as the primary source of taxpayer related information. The database for this system resides in a different computer to those used in the ACS. The two systems are not compatible with each other. There were several consequences of this lack of integration.

In discussions with taxpayers regarding their tax liability, Contact employees might be informed that certain payments or activities had recently occurred. In these situations, the Contact employee would experience embarrassment because one hand of the IRS appeared not to know what the other hand was doing. This would occur because information on the IDRS system was obtained at the Service Centers. Any time delays incurred at the Service Centers in updating IDRS information was compounded by the time delays in inputting this information into the ACS computer systems.

In such situations, the Contact employee had to walk over to an IDRS terminal and examine the records to verify the taxpayer's information. This process required a knowledge of IDRS procedures and access authority, which not all ACS employees had. Some sites instituted a procedure whereby ACS employees filled out a form requesting such IDRS verification and submitted these forms to a designated IDRS operator. These procedures and the lack of interface between the two systems added to the delay in maintaining current ACS records.

A second system procedure also affected the operation of the ACS sites. The procedure for assigning cases to a contact employee for follow-up were on a next-case basis. As soon as the contact employees pressed the 'next case' button on their terminal, the case with the highest priority was directed towards their workstation. No procedures existed to assign the more difficult cases to the more experienced employees. A brand new employee might be assigned an extremely complex case while veteran employees were working low dollar, simple cases. The inequities caused by this procedure are currently being addressed by the new 'team' approach discussed in Chapter 7.

Although it was originally planned as a paperless office, the ACS operation resulted in a reduction of paper rather than a totally paperless environment. The large volume of incoming correspondence, IRS and ACS rules, procedures, guidelines and memoranda soon filled several filing cabinets at all sites. What did disappear were the taxpayer account files which had threatened to overwhelm the former Collection Office Function.

An administrative issue related to personnel activities was that of shiftwork. Due to the nature of its activities, the ACS needed to contact taxpayers directly. During the hours of the traditional work day this contact was not always possible. It was often necessary to call taxpayers during the evening and weekend hours. In theory, ACS call sites were to be in operation for a 60-hour work week. Evening and weekend hours proved to be not particularly popular with many of the ACS employees or managers due to family considerations, transportation problems, personal security concerns and other restrictions on their life-style. In response, several sites maintained only a limited staff during evenings and weekends while other sites voluntarily limited work hours to traditional schedules.

The identification of site locations created much heated discussion. The issues revolved around the relative merits of an urban *vs* a suburban location of the site. On the side of those who promoted the urban side of the argument were such factors as the existence of an established workforce, lower costs (since remodelling was less expensive than construction), the availability of public transportation, and concentrated sources of employees. The arguments in favor of a suburban location tended to emphasize the desire to tap alternative labor sources (i.e., suburban housewives), the need to construct facilities from scratch specifically designed for the ACS, and aesthetic values of a suburban location. The issue of urban-suburban locations was resolved at the National Office level by allowing each region to make its own decisions regarding where to place the site.

An administrative issue that arose early in the implementation may be unique to the IRS as an organization. In selecting managers for the ACS sites, the emphasis was placed on managerial potential rather than extensive knowledge of ACS operations and procedures. Many first-line managers were drawn from other branches of Collection and the IRS. In several cases, the managers possessed less technical knowledge of the system than the employees they managed. To alleviate the potential problems that this situation might create, a large number of Revenue Officers from the Field Collection Branch (see Chapter 2 for a description of this position) were temporarily detailed to the ACS to act as on-the-job coaches during implementation. Frequently, employees would approach the Revenue Officer with a question rather than their immediate supervisor. The relationship between ACS employees and the Revenue Officers developed into a very positive, supportive one. Revenue Officers, for their part, enjoyed the situation since they did not have the administrative responsibilities of management, they had the respect of employees, and were taking a break from their normal routine.

This approach, however, proved to have negative consequences. The establishment of bonds of respect between employees and Revenue Officers tended to diminish those between the employees and their immediate supervisor. At times, employees would approach both their coach and their supervisor with the same question and get two different answers. This was especially true during the initial phases since procedures had not been established for many of the situations that employees would encounter. Decisions in these cases were often made on the basis of judgment and experience rather than established guidelines.

With the withdrawal of the Revenue Officers following the end of their detail, the relationship between employees and their supervisor was often ambiguous. It took some time and a great deal of effort to establish a normal supervisor–subordinate relationship between these two groups.

Supervisors, for their part, were also caught up in many of the organizational and administrative issues that existed during the implementation. One of the most important of these was the use of the Management Information System (MIS) which had been designed to provide them with timely, comprehensive, and detailed information on the status of relevant ACS and employees operations. In this situation, it proved to be too much of a good thing.

Part of the problem with the MIS was due to the training managers received on its use. This proved to be somewhat cursory. Another part of the problem was that the information was, in many cases, too detailed and too voluminous. Many of the reports that were available to managers were at a micro level when, in fact, the managers needed summary level reports. Much of the information was of only tangential importance.

The functionality of such information was compounded by other demands made upon the manager's time. In addition to such normal administrative managerial responsibilities related to personnel activities, managers were expected to monitor employees' performance several times a week, complete a large number of productivity report forms, keep abreast of almost daily changes in system operations, counsel employees, attend management briefings, brief employees, etc. Several sites required managers to perform duties ancillary to their job as manager, for example, building security responsibilities.

Taken together, the functional limitations of the MIS system and significant demands on the manager's time and attention resulted in two consequences. First, many managers made less use of the MIS than expected. Second, reports of managerial burnout began to be received from several sites. Upon investigation, the reports of burnout proved to be more a case of overload and conflicting responsibilities than the classic symptoms usually attributed to managerial burnout.

Two forces acted to ameliorate these conditions. For the most part, the frantic activities and ambiguities that accompanied the initial stages of implementation tended to decrease as the system 'settled in'. The number of and conflict among different managerial responsibilities tended to even out over time. The reports issued by the MIS were also revamped to be more consistent with the informational needs of the manager. While the demands made on managers remain high, the situation appears to have stabilized.

In summary, the implementation of the automated systems put a great strain on the administrative and organizational structures designed to support the system. Where procedures had been established, they tended to be interpreted and modified according to local conditions. In those areas where guidance did not exist or could not be obtained in a timely fashion, local initiatives were devised. Although many local variations exist, major differences in site operations and procedures were later standardized through national office directives. Many of the procedures that did not work as well as planned were modified or adapted over time.

Section III

LESSONS LEARNED

Chapter 11
Automation and Leadership

1. The role of the supervisor

The need for good management is not unique to an automated office. Managing people effectively is the foundation of all productive organizations. The need for these skills in an automated environment becomes, perhaps, more apparent because of the increased speed, complexity and consequences resulting from the automation of manual systems. For example, under a manual system an individual using an incorrect method of filing information might affect hundreds of cases before the error is detected. The same individual, under an automated system, might well misfile thousands of cases with little hope of easy retrieval and correction.

Automated systems tend to magnify the consequences of good or poor management practices that existed under non-automated situations, in addition to alleviating old ones and creating new ones. Relatively minor problems that were annoying to employees under manual systems may well become major irritants under automated conditions. Dissatisfied employees express that dissatisfaction in many ways. Probably the most common means of verbal expression is complaining.

In dealing with the day-to-day issues that arise, it is easy to lose sight of the fact that employee complaints can be easily misunderstood by managers. Two types of complaint patterns have appeared in our studies. In one type, a continual stream of 'small', low-level problems are raised in a sequential fashion. Considered individually, each problem or complaint appears trivial but the pattern of fire fighting persists over an extended period of time. The second type of pattern occurs where the employees focus as a group on what they consider a major problem. Emotions tend to run high and appear to be contagious.

The typical manager will attempt to deal with each problem as it arises only to be confronted with another, different, issue. Despite the obvious fact that such a band-aid approach to problems usually fails, few managers recognize that the real problem is not the obvious issue at hand but, rather, the underlying motivations and satisfaction levels of the employees. The existence of long-term systematic complaining is usually a symptom of a much broader based dissatisfaction. In an organization where employees are highly motivated and satisfied, the level and intensity of complaining are negligible. A 'make do' attitude exists even under adverse conditions. Problems are treated as issues and dealt with openly among the employees and supervisor. Few incidents of finger pointing and blame are evident.

At the opposite end of the satisfaction and motivation continuum, employees see problems at every turn and take every opportunity to express their dissatisfaction to each other and their manager. Rather than deal with the underlying issues, it is easier to find fault with the equipment, the environment or the organizational structure/procedures. This type of atmosphere, once established, tends to be self-perpetuating. Even dealing with the immediate source of irritation effectively does not address the underlying issues. More frequently, the inability to immediately and effectively deal with the source of irritation contributes to increasing levels of frustration for both the manager and the employee. Not surprisingly, few of the individuals involved in this cycle are aware of the real reasons for its existence. It is all too easy to emphasize the immediate, concrete problem and ignore the more difficult and long-term issues of managerial style and group relationships.

The bulk of this chapter attempts to describe some of the behaviors that employees and managers have identified as promoting a positive and productive organizational climate. The characteristics described are those which were identified in group and individual interviews over a period of years, geographical locations and functional areas.

2. General considerations in managing

Along with their peers in the work group, the manager is the primary personal contact with the organization. The rest of the organization is identified as an anonymous 'they'. If the group perceives the manager as a member of the organizational bureacracy rather than as part of the group, the supervisor will lose the trust and respect essential for a leadership role. With the loss of trust and respect, the supervisor must resort to power based upon authority to direct individual activities. When this situation occurs, an adversarial relationship frequently occurs. Once established, this relationship between the supervisor and the work group is extremely difficult to alter. New members are quickly inculcated with the 'appropriate' attitudes and values regarding the supervisor. The more resistance to supervisory authority the employees display, the more likely it is that the supervisor will rely on authority, rules and regulations to bring the employees back into line. It becomes a vicious cycle reflected by hardened positions on both sides.

The individual employee resents excessive reliance on formal authority and will resist in a combination of active and passive modes. The net result of this inevitably resulted in such typical reactions as:

Psychological withdrawal and apathy.
"It doesn't bother me one way or the other."

Pessimistic, doom-and-gloom attitudes.
"Take my word for it, things will go from bad to worse."

Minimal levels of effort.
"That's good enough for government work."

Sticking precisely to the rule book.
"That's not in my job description."

Formal complaints over issues that could have been settled informally.
"Put it in writing if you want any action."

Deriding those who are promoted or advanced.
"Mary knows how to buddy up to the boss. That's why she got the job."

Placing blame on others for their lot.
"If only (group or individual) would get their act together, life would be lot easier around here."

Stereotyping others.
"The only thing that moves in Personnel is the clock."

Sabotage.
"Hey, look what happens to the printer if a pencil accidently falls in it."

These type of comments, or their absence, are indicative of the organizational and work group climate. The climate, among other factors, is a result of the managerial style. An attentive observer needs only a few hours with a work group to get a feel for what type of leader the supervisor is.

Using the group interview technique described in Chapter 9 in conjunction with individual interviews and a literature review, three major categories of managerial behavior appear to form the bases for highly effective work groups. These three categories are 'Expectations and performance', 'Flexibility and understanding' and 'Decision making'. The distinction among the categories is arbitrary. Many of the behaviors listed under one heading could just as easily fit into another category.

Expectations and performance

Few employees felt motivated by tasks which required little challenge and respect. Satisfaction with work is achieved best when the employee meets or exceeds high standards and expectations. Setting standards significantly above or below employee abilities and motivations over a long period of time will significantly affect turnover, job satisfaction and productivity. Few people object to infrequent levels of very high or very low demand. It is the chronically high or low levels of performance expectations that are detrimental to satisfaction and productivity.

There is an inherent desire on the part of employees to do their job well and be recognized for their efforts beyond average levels of performance. Few people like to fail at their job. Employees' perceptions of meeting, exceeding or failing job expectations are a function of the feedback that people get from the work itself, their co-workers and their supervisor. Rewarding people for minimal or average performance often acts as a demotivator for the employee and other workers. Some of the most typical managerial behaviors and employee reactions that fall in the area of rewards and expectations that we have observed or that have been identified by employees are:

1. Making frequent exceptions to reasonable expectations will cause those exceptions to be expected.

2. Reward outstanding and punish poor performance as soon as possible. The longer the delay between performance and reward, the less the effect of the reward or punishment.

3. Expectations for performance should increase by 5–10% per unit of time. Once a task has been mastered to the point where potential increases in productivity are limited, evaluate the task itself in terms of redesign.

4. Performance comparisons should be against some objective standard that is mutually agreed upon. When no objective standard exists, performance should be measured in terms of the activities necessary to achieve results, not the results themselves.

5. It is better to have outcome standards mutually agreed upon and let the employee deal with the means to achieve those outcomes. Specifying not only the results but the means to achieve those results leaves the employee little discretion, personal responsibility or latitude for creativity.

6. Share the dirty work among employees and, occasionally, the supervisor. The boss who 'rolls up his sleeves' to take part in employee tasks when the situation allows is valued by the employees as a team member. It also allows the manager to appreciate what the actual tasks require of employees.

7. The manager should act as an advocate for the group as well as the individual employee within the organizational hierarchy. Employees who see their manager 'fighting' for them are much more likely to fight for their manager in terms of productivity.

8. Never play favorites except in terms of employee ability to get the job done. Even then, it should be clear to all group members that one individual is more competent in that specific area than the others.

9. A supervisor should be friendly but not friends with his or her employees. To become friends with a member or members of the group sets up expectations on the member's part for special considerations. When work decisions have to be made the same standards should be applied to everyone, regardless of personal feelings.

10. For important expectations that absolutely have to be met, build in back-up systems for their accomplishment. Assume that the initial plan to get the task done misfires. What alternative ways and resources are available?

11. When expectations cannot be met, the supervisor should choose his or her fights carefully. It pays to fight for things that matter and that can be changed. Getting into fights about matters such as individual dress standards when it has no effect on performance is an exercise in futility.

12. Take the long-term view of things when dealing with employees. The performance expectations you have for an employee should not focus entirely on current or near-term activities but also on the long-term, career, expectations. A manager should assume that he or she is training future managers for the organization.

Flexibility and understanding

A manager who works strictly 'by the book', in a sense, may represent the ideal first-line supervisor. Certainly, rules, procedures and regulations exist for a definite organizational purpose. They reduce the variability and ambiguity in situations. Additionally, if based upon successful experience or well thought-out planning, they ensure that the most beneficial alternatives are used. A manager who finds it difficult to improvise, make exceptions or recognize new situations will have great difficulty managing an effective group in all but a few highly routinized situations. Rules, standards, procedures, etc. should be viewed as guidelines. The manager who exhibits a flexible attitude is seen as a person who 'gives a damn' and has a personal interest in his or her employees. As in the previous section, the assignment of behaviors to this category rather than another is somewhat arbitrary on the author's part.

1. Take a 'we' approach to problems. An employee- or manager-centered problem is rarely the result of one individual. A personnel problem exists when two or more people interact. Both must recognize that they contribute to the problem even if not in equal share. All of those involved should take some responsibility for improving the situation.

2. It's acceptable for managers to admit they are human and make mistakes. The manager who 'never does anything wrong' has a group of employees just waiting to

catch him or her in a mistake which they will never let be forgotten. Making fun of yourself also shows that you are human.

3. Little things mean a lot. The manager who remembers an employee's birthday, children's graduation, and says 'thank you' sincerely gives employees a motivational boost that will last a lot longer than any pay raise.

4. Communicate with the employees, even on non-essential information. In the absence of information employees will fill gaps with rumor and guesses. For the most part, these rumors will not reflect positively on the manager or the organization. Explaining why something was or was not done defuses most of these rumors and builds trust.

5. Never lie. Employees who catch a manager in a lie will rarely, if ever forgive him. Years of trust go down the drain and will affect all future relationships.

6. Give in to employees on non-critical issues. Even if it is against your better judgment, let employees make the decisions from time to time when failing is not critical. If the employee is right, you've acquired a valuable insight into your employee's competence. If the employee is wrong, he or she will trust your judgment more, providing you use the failure as a training exercise.

7. Draw people out. Psychologically, how you say something is as important as what you say. Learn to ask open-ended, non-threatening questions. This is easy to do but hard to learn. The easiest way is to ask for people's opinion regarding an issue rather than ask for a 'yes' or 'no' type of answer.

8. Go crazy every now and then. Total predictability is boring. On a more or less random basis, do something unexpected like put all the phones on hold and take a group coffee break away from the office.

9. You can't argue rationally with people who are in a very emotional state. You have to deal with the emotion first before you can solve the problem. This doesn't mean that you should let the employee cool off first. The easiest way to deal with the emotion is to define it from your point of view. For instance, saying to an employee "You're really unhappy about that decision, aren't you?" rather than debating the decision itself defuses much of hostility.

10. Criticize people in private. Dressing down an employee in public loses a chance of improving his or her performance. The humiliation the person feels will last a lot longer than whatever it was that went wrong. The best place to argue with an employee is on neutral ground, like a cafeteria, and not in your or his office.

11. Never deride an employee behind his or her back. Eventually it will get back to them, usually in a distorted fashion.

12. Don't leave employees hanging on. Unjustified procrastination in making a decision or dealing with a situation says to employees that you don't know, don't care, or are afraid of dealing with the issue.

13. Symbolism is often more important than actual events. Awarding a person a title promotion, a minimal cash award, a logo key chain, etc. often takes on a psychological meaning far beyond its cash value.

14. It is frequently more effective to give group-based performance feedback than individual. Group members are much more aware of who contributes to productivity. Giving group feedback immediately creates peer pressure on low performers. Such pressure is frequently more effective than managerial threats.

15. Let the group, whenever possible, deal with goals and issues and challenge them to deal with them. Trying to set a coffee break schedule to satisfy everyone is doomed to failure. The manager should establish the criteria (e.g., at least two people available to answer the phones) and let the group decide how to implement it.

Decision making

The last of the three areas of behavior cited by employees and managers is that of decision making. In a very broad sense, a manager who is rated high in this area is viewed by his or her employees as competently performing the technical and administrative functions of the position. The behaviors that have been frequently been cited by employees as important in this area are:

1. Take a balanced view of performance. Managers frequently tend to overreact to a single event rather than take a broader, more balanced view. Disciplining an employee should not be done for an isolated act. Patterns and trends should be the basis of any action.

2. Show respect for the professional skills of employees. Trust employees to use their own judgment and insight whenever possible. The real role of the manager should be that of a problem solver (when it is beyond the skills, knowledge or abilities of employees). In all other cases, the supervisor should act as a facilitator.

3. Don't show off to employees. In many situations, the manager can do the job better than the employee. It's acceptable to demonstrate a task for an employee but don't do it for them. The worst case occurs when the manager tries to impress an employee but fails to do it as well.

4. Give clear feedback and directions. Nothing is as frustrating to employees as being told that something was not done correctly and being told to "fix it". Employees frequently think that managers who say "I'll know what I want when I see it" really don't know what they want.

5. When decisions are made that directly or indirectly affect employees, the rationale behind the decision should be explained. Employees will frequently accept even unpopular decisions if they understand the need or logic underlying that decision. It you can't explain the logic behind a decision you're in trouble to begin with.

6. Promise only what you can personally deliver on. Employees frequently make inferences far beyond what is actually said. If you imply that a promotion recommendation will follow some performance, the employee will remember only that a promotion will follow.

7. Maintain a balance between wcrk and other activities. Work, while an important part of life, is only a part. People have an existence beyond work and there are limits to what long-term demands the organization can make on employees.

8. Know where to draw the line in taking action. Unless it affects on-the-job performance, the manager has no legitimate right to discuss a problem with the employee. However well-intentioned, discussions relating to alcoholism, drug abuse, sexual promiscuity, etc. are inappropriate for the manager to initiate.

9. Require employees to act as instructors from time to time. Employees, because of their daily involvement with the job, know a lot more than training manuals. By having skilled employees take the responsibility to train others, everybody benefits.

10. Find reasons to delegate, not reasons to avoid it. One of the paradoxes of management is that the closer the manager tries to control everything that goes on in his or her section, the less control he or she actually has. If the manager needs to be consulted on every issue that arises, he or she will be continually dealing with many trivial requests. Little time will be available for more difficult issues or long-range concerns. Lack of delegation also creates a feeling of paternalism among employees, who resent being treated as children.

11. Don't let the exception determine the rule. Even in the best of managerial–subordinate relationships, the freedom, responsibility and trust that exists will occasionally

be abused. A common managerial reaction when this occurs is to increase control and limit delegation to avoid a repetition of the event. Treat failures as a learning experience rather than a punitive one. Don't punish the 99% success rate for the 1% failures.

12. Don't try to be a psychologist, be a manager. Most people are not completely aware of all the reasons why they act the way they do. Motivation is a complex interaction of important and minimally important needs, long- and short-term goals, unique events and general conditions, past experience and future expectations. Few behaviors are motivated by a single cause, and they are not exactly the same from day to day. Rather than placing emphasis on why an individual did or did not do something, it is more effective to concentrate on the behaviors that led up to an event and the consequences of those behaviors.

To the typical employee, these specific behaviors represent the underlying attitude of the manager towards the employee and the work group. The manager, to be maximally effective, must be respected and trusted by the employees. This appears to be true whether the work environment is manual or automated and with little regard to the work content. There are, of course, several important differences in managing an automated office. It is to these that we now turn.

3. *Managing in an automated environment*

Automation and computerization are introduced into a total work system in order to increase productivity. Increased productivity should not be simply equated with increased efficiency. Once the new technology is in place and managers and employees have sufficient training and experience, there is usually a marked increase in productivity: more is produced at a lower cost. It is easy, at that point, to lose sight of the long-term need to sustain and improve productivity. Automating tasks and processes provides an initial spurt in productivity, but effective management must be concerned with how the human element can build upon and enhance the technological improvements. Technologically generated productivity improvements can deteriorate when managers are not attentive to the human aspects of work.

Learning to manage in an automated environment does not entail unlearning the basic principles of sound and effective management. The points raised in the first part of this chapter are still valid. However, they apply in significantly different ways. These differences are caused by the very nature of automation and how it affects the way in which individuals interact with each other and their work. Management which fails to recognize and adjust to these differences will not be effective.

Among the significant differences in managing people in an automated environment are:

Physical demands

Employees in a computerized system are more deskbound; they have less occasion to get up and move around the workplace and so less physical variety in their movements, postures and environment. Sitting at a VDT for an extended period of time imposes high demands upon an employee's visual and musculoskeletal systems. Prolonged periods of restricted movement often create a degree of fatigue and tension. Eyestrain, which may result in head- or neck-aches, often occurs as a result of continually refocusing from paper documents to screen displays. Backstrain, leg cramps and similar muscle reactions result from maintaining an incorrect or poorly supported posture or seating position.

In order to capitalize on the efficiency of the automated system, breaks and rest periods are more strictly regulated. Even when the work process is not computer-paced, there still exists an emphasis on speed of entry and execution for the positions associated with the automated systems.

Mental demands

Work in a computerized environment demands more attention than the same type of work in a non-automated environment. Employees, for extended periods of time, focus on display screen information, paper documents, entry information and keyboard functions, in addition to other required tasks. A high degree of accuracy is often required since errors may easily go undetected. Although opportunities for errors on a large scale may be limited, a small error may have a ripple effect, creating successively larger problems.

The restrictive nature of computerized data entry and retrieval often contributes to the problem of boredom which, in some ways, is the mental parallel of physical fatigue. The repetitive nature of some tasks allows the operator to withdraw psychologically from the task. Task attention wanders and when a variation or anomaly in the routine occurs, an error frequently occurs.

Social relationships

An automated environment imposes distinctive social constraints upon employees. Interpersonal relations contribute directly to job satisfaction, group morale, motivation and information transfer. However, both in terms of physical layout, equipment constraints and job structure, work in an automated environment tends to isolate employees from each other and their immediate supervisor. Employees in an automated environment, on the whole, tend to work more independently of each other and their manager than they did in a manual systems environment. In addition to the increased focus on the display and entry medium, work cubicles and sound and sight barriers between employees heighten the sense of isolation when compared to a non-automated environment.

The atmosphere within an automated environment is much more subdued and quieter than its non-automated counterpart. There are several reasons for this: many 'products' are electronically displayed; printers are sound-proofed; external environmental noises are reduced by the necessity of air conditioning and non-opening windows; and carpeting and soundproofing also contribute to the reduced levels of internal sound.

The impact of social isolation on people appears to be more acutely felt by employees who previously worked in a non-automated environment. In general, people resent not being able to communicate occasionally with their co-workers, especially when the task requires long periods of individual focused attention. From a managerial point of view, the restriction of interpersonal contact may seem to be highly desirable. The assumption is that social interaction is not task related and thus a waste of time. It has been our observation that this is not the case; most interaction deals with task or organizationally related questions. When personal, non-work related issues do arise, it can be argued that these discussions act as a pressure valve and a psychological break from the work. As such, it tends to break down the psychological isolation caused by task accomplishment. It is the abuse of social interaction, not the interaction itself, that reduces effectiveness.

Impact on skills

Competence, pride and personal confidence in one's skills are factors in job satisfaction and motivation. The automated work environment dictates its own relevant skills and may usurp tasks or processes which were once accomplished through human skills. Workers who come from a non-automated environment (particularly if they performed the same sort of tasks there) may feel that they have been de-skilled in their new environment. An activity that may have been, at one time, a source of pride with a sense of accomplishment may be regulated to a computerized routine. A secretary who, for example, prided herself on her skills in maintaining an orderly, up-to-date set of files may have these hard-earned abilities replaced and improved upon by a computerized database program.

Second, the use of computer routines frequently reduces the amount of ingenuity, insight, creativity and individuality that was the hallmark of the competent employee. An automated work environment tends to narrow the scope of employee decision-making to data entry and command execution via a keyboard. Limiting employee discretion has the advantage of uniformity but, in many cases, the employee loses the feeling of responsibility and accomplishment that is gained from years of experience or insight in making an informed decision. Automation of an activity requires that the total process be analyzed and broken down into its sequential component parts. In a highly rational format, options, decision points, flow, information input and outcomes are defined. The net result of this analytic procedure and its routinization is to leave little room for those human factors which define the superior employee: creativity, flexibility, insight. The automation of routine activities might free the executive or manager to pursue more challenging tasks but, for the average rank and file employee, the presence of these alternative activities is not an everyday event. While the executive might feel liberated, many employees feel controlled. Instead of controlling the work process, many employees feel controlled by it. This reaction is particularly evident when the employees are restricted in their physical mobility, their work is machine-paced, or they are subject to close monitoring by supervisors and computer systems.

Third, the process of automation, by requiring a rational sub-division and fractionalization of work activities, appears to restrict employees perceptions of the total work process. If the automation effort results in job simplification rather than job enrichment, employees feel that they have become more remote from the actual work of the organization. In a very real sense, this 'task' isolation is similar to the social isolation previously discussed.

4. Performance characteristics

Quantitative standards

Computers are machines that thrive on numbers. They carry out tasks via commands, data and programs that are essentially numerical. The very engineering of an automated work process is biased in favor of identifying productivity in terms of time and quantity. Since it tends to limit the variability of employee behavior, automation promotes categorizing productivity in terms of such criteria.

Productivity criteria and measurement, which are difficult or impossible to gather using non-automated work systems, are effortlessly gathered in an automated environment. The computer becomes, at the same time, both an instrument of productivity and control. This dual function may instill ambivalent feelings in employees about automation: one and the same instrument is supposedly an aid in accomplishing the job, yet can also 'spy' on the employee.

One of the consequences of automation is that it can give some very real 'hard' indicators of system and individual performance. The availability of such information, however, easily leads to an overemphasis on numerical indices. It becomes difficult to keep sight of the goals of the unit when emphasis is placed on meeting quantitative goals. The danger exists that quantitative assessment will become an end in itself to the detriment of the real organizational goals—producing goods or services.

Managers who tend to rely entirely on objective standards of performance and productivity are in danger of missing many of the other factors which are critical to human performance. An employee, for example, who is frequently interrupted by questions from co-workers may see his or her own productivity levels decrease while those of others increase. The apparent requirements for face-to-face interaction in an automated environment decrease due to the generation of quantitative information relevant to performance. The real need for such personal interaction and communication actually increases.

Increased stress

Both because of what the automated work environment demands from the workers and what it takes away from them, stress is different, and perhaps greater, than in a non-automated environment in which the same or similar tasks are performed. Such stress is induced by the factors we have just described: task simplification, repetition, machine-pacing, increased monitoring and control of performance, etc.

Stress can never be eliminated from the workplace; nor should it be. Although the reaction to stress is a very individualistic phenomenon, a moderate amount promotes an attentive and motivated attitude. Organizations frequently use goal setting, performance awards and group competition to induce stress. High-level or chronic stress, on the other hand, causes degradations in performance, morale and job satisfaction. Particularly during the transition period from a manual to an automated environment, the pervasiveness of stress is multiplied by the need to learn new skills, debug the system and establish new work relationships, compounded by the necessity of productivity goals.

Increased complexity

Automating tasks and processes has the immediate effect of simplifying the work system by generally reducing the number of people needed to accomplish a task. However, this is a deceptive simplification. In actuality, system components, including the human element, become more dependent upon each other. Improvements, changes or breakdowns in one part of the system frequently have a direct or indirect effect on other parts. While attention is frequently paid to the direct impact of any change, organizational relationships are so interdependent that a 'ripple' effect can often cause more indirect changes than anticipated. Some of these effects cannot always be foreseen. Improvements in one part of the system can result in problems in another part.

Increased opportunities

It is neither possible nor desirable to plan for every contingency. Dealing with the increased complexity of automated systems requires planned flexibility. Flexibility in an automated environment requires planning for back-up systems, a trial period, sensitivity to clients and employees, and a willingness to modify, augment or drop activities.

The period of new technology introduction is ideal for experimenting. Since no system can be planned in every detail, the flexibility to experiment is at its maximum during the implementation stage. If risks are to be taken, this is the point at which they should be. The nature of the implementation period is that of a trial or test to establish the parameters of the system and the people who run it. It is a time of learning which should have as its focus new ways of doing things. Frequently, the attitude of 'let's wait until the dust settles until we try anything else' is counterproductive to long-term system success. This does not mean that anarchy should rule in the workplace. Far from it Any changes in the systems should be thought out to the maximum extent possible. What it does mean is that creativity and intuition should be brought to bear by all those involved in addressing issues, problems and opportunities that arise during the introduction period.

Existing problems may be magnified

One of the most common, if unspoken, beliefs of new system planners is that many, if not all, of the problems that existed under the manual system of work will be resolved with the advent of the automated system. This may be true for activities that are to be eliminated or radically altered. However, problems that existed under a manual system can often be exacerbated under an automated system; they don't just simply disappear. Poor management–employee relations become even poorer, lack of incentives become dis-incentives, poor planning is more obvious, limited resources become more restrictive, poor communication can become disaster.

Automation is not a panacea for all organizational problems. Unless eliminated by automation, problems that existed under non-automated conditions will tend to increase under an automated process. Cracks tend to become fissures when the stress of automation is introduced.

5. *Organizational culture and management*

Several aspects of the organizational culture frequently promote a transfer of managerial style from the manual to the automated environment. A pervasive philosophy is 'What worked well then will work well now'. Few organizations recognize the differences in management behavior that are required of an automated system. If any recognition of the differences is made, the organization typically assumes that some form of management training will suffice. Management training in the technical aspects of the systems does nothing to affect management style. Supervisory style training, if it exists at all, is usually embedded within the context of technical training and consists of a series of lectures on the virtues of theory X or theory Y management attitudes. The assumptions underlying this approach are several:

You can change attitudes by education.
Attitude change preceeds behavioral change.

> The best way to change attitudes is in a classroom, surrounded by your peers. Give people the theory (should do's and don'ts) and they will know how and when to apply the concept in the real world.

None of these assumptions, of course, hold up in the real world. Management training in human resource management, except in unusual cases, accomplishes little. On-the-job experience between employees and supervisors has the most direct and significant effect on attitudes, not the other way around. In our experience, a manager who sees himself as an effective leader of people when supervising a manually accomplished task will attempt to apply those same skills in an automated environment. Lack of managerial success in the automated environment is frequently responded to by defensiveness and attempts to 'better' apply the principles by which they achieved success in the non-automated environment. Few, if any managers, recognize that a change in managerial style, not an embellishment of previous styles, is most effective.

Another common organizational aspect that promotes ineffective management is the philosophy of equating technical skills with managerial skills. This often unspoken belief assumes that if an individual exhibits technical competence in performing a task, he or she is likely to be the best candidate for managing a group of people who perform that task. While it is highly desirable to have a manager who is competent in both the technical and managerial roles, it is our experience that human resources skills are the most important of the two for effective task accomplishment.

A final generalization concerning common organizational philosophies that tend to restrict effective management has been termed 're-inventing the wheel'. It is not uncommon for management to be somewhat faddish in its approach to organizational initiatives. Management-by-objectives, socio-technical systems, open-systems approaches, strategic planning, transactional analysis, zero-based budgeting, and so on, have all been 'hot' topics at one time or another. When, as is natural, the publicity and the practice don't coincide, the new approach is discarded and the next 'magic bullet' is advanced. Few programs are total failures or total successes. There exists, however, a tendency to throw the baby out with the bath water. A change in management approach needs to be taken from a success/failure orientation to a more iterative style of attempting programs to improve human resources management.

6. Conclusions

This chapter has attempted to identify some of the issues, problems and opportunities that management will face in the transition from a manual to an automated environment. The major conclusion to be reached in this context is that automation is not a magic pill that, once swallowed, will free managers from dealing with the foibles of the people and systems they manage. In reality, good management skills are remarkably similar in both an automated or a manual system. Automation tends to place additional emphasis on good managerial skills for dealing with employees, not less. The success or failure of automation ultimately does not rest on hardware or software. People can make the best designed systems a total failure. Alternatively, they can make even the most poorly designed systems a success. Walton (1980) points this out quite clearly: "In the absence of commitment strong enough to sustain high performance, carelessness, inattention, and alienation will undermine even the most sophisticated technology".

Chapter 12
Motivation and Productivity

Introduction

Little doubt exists that technological advances have the potential for contributing to improved productivity in the workplace. In the final analysis, however, it is the individual and group effort in using the new work technologies that accounts for both the greatest costs and potential for productivity gains. The best designed systems are doomed to failure when met with human resistance. In contrast, even the most poorly designed systems have a significant chance of success if employees are motivated to that end. The difference between the two reactions to new technologies revolves around the issues of employee commitment and motivation. Committed and motivated employees are the consequence of internal and external rewards. Employees are motivated and committed to the extent that organizational rewards are consistent with their values, attitudes and needs.

The initial start-up period for the ACS created a unique atmosphere among the employees, managers and executives. The shared experience of participating in the design and implementation of a major change within the IRS was clearly a motivating factor of immense force. The effort required to get the system into operation required a high-level effort over an extended period of time. With few complaints, both employees and managers alike made Herculean efforts when required. In spite of the difficulties encountered in setting up the ACS system, morale and productivity were high. The expectation of contributing to a major change in work procedures, the lack of established procedures, the reduction in formality and the 'novelty' effect all contributed to high levels of dedication. The ability to maintain these high levels of commitment over an extended period of time proved to be another matter altogether.

Much of the previous discussion highlights the fact that satisfaction and individual productivity levels either stabilized or tended to decrease over time. As the new became the familiar and flexibility became regimen, the ACS tended to acquire some of the organizational and managerial styles and problems that existed prior to automation.

The recurrence of old patterns and values following the relative freedom and comradeship of the implementation period was resented by the employees. As a consequence, turnover rose over the course of time, union grievances increased at several sites, and the general level of complaining, blame placing and finger pointing grew.

After several site visits the pattern of relationships between time (implementation period to stabilization) and employee motivational level became clear. It seemed clear that the major human resources challenge of an automated system was not the implementation but the maintenance phase. As a federal agency, the Internal Revenue Service's ability to maintain high levels of motivation over extended periods of time through reward systems are subject to federal laws, rules and regulations. In contrast to private sector organizations, federal reward systems are typically bureaucratic, cumbersome, paperbound and time-consuming.

In the process of visiting several sites a number of requests were made for a prescribed series of activities to maintain or improve the motivation of employees. Rather than deal with these requests on a site-by-site basis, a variety of activities were written up and distributed among those call sites that expressed an interest in this area. These suggestions for maintaining high levels of motivation are described in the remainder of this chapter. Without on-site support and guidance for these suggestions few, to the best of our knowledge, were fully implemented. The suggestions remain as a useful resource for planning new project implementation.

1. Background

Many organizations and managers assume that commitment and motivation occur in direct proportion to salary and benefits, including work environment, career progression, retirement plans, recreational facilities, etc. The importance of these factors appears to be most critical in the decisions to initially join the organization or to leave the organization for a 'better' job. The determinants of day-to-day commitment and motivation do not appear to be tied to these 'extrinsic' factors (Herzberg *et al.*, 1959). Few employees put in extra effort because of improvements in the retirement system. A frequent error of management is to assume that a pay increase or similar tangible benefit will improve commitment. Such extrinsic rewards as money can achieve increases in specific behaviors over a short period of time but their effects decline over extended periods. Offer an employee a 10% bonus to finish a job ahead of time and he will be motivated to comply. The commitment is to the money, not the job or organization. On the next similar assignment he will expect another offer of a bonus. In fact, the lack of a bonus may well act as a demotivator since the expectation of such rewards has been increased.

In a similar sense, the effects of money as a motivator are relative. The value to an employee of a reward is not judged in isolation. Its value is made in comparison with some reference point. At least three comparative standards are used by people to assess the value to them of rewards. The first reference point is vertical, the second horizontal, and the third may be termed market value.

Vertical comparison refers to the assessment of rewards against prior rewards. A bonus of $500 may seem small compared to a previous bonus of $750. A horizontal comparison refers to an assessment against the rewards given to co-workers in similar circumstances. Here again, a $500 bonus loses its motivating value when a co-worker receives a $600 bonus. Finally, individuals make comparisons against an assessment of the market worth of their efforts. The motivating potential of a $500 increase in salary is minimal if comparable positions in other organizations receive often higher salaries.

A common theme throughout each of these comparisons is the psychological evaluation of the comparison and the fact that rewards have the potential for demotivating individuals as well as providing incentives for performance.

Authentic motivation and commitment must be based upon the value system of the individuals and the nature of their relationship to the work and the organization. Money will buy mercenaries, involvement buys commitment.

Walton (1980) identifies four levels of employee involvement and commitment:

1. Moral commitment is based upon the belief that what one is doing is morally correct.
2. Spontaneous-expressive commitment is based upon the feeling that what one is doing is enjoyable and gratifying.

3. Calculative commitment involves a trade-off between individual effort and organizational pay for that effort.
4. Alienative or coercive commitment involves the use of fear and/or punishment to achieve compliance.

Etzioni (1961) says that organizations maintain a value system orientation similar to that of the individual. A prison, for example, would be representative of an alienative/coercive orientation, a sales organization would be representative of an organization fostering calculative commitment, a circus might represent a spontaneous-expressive commitment, and a charity fund organization could be taken as representative of a moral commitment. To the degree that individual commitments and organizational values are consistent with each other, high job satisfaction and motivation are more likely to result.

Most organizational incentive systems are, in practice, geared to promote calculative (i.e., pay for performance) or alienative commitment (i.e., follow the rules or face the consequences). This practice is in contrast to the usual official company policy which espouses the need for moral commitment of employees to the goals of the company. It is possible for almost any job to be justified on the basis of moral righteousness and contribution to the public good. The fact that there is frequently a difference between the official image and unofficial practices and values in the organizational culture is not lost on employees.

It is possible to buy people to perform undesirable tasks in exchange for high pay or other extrinsic rewards. Such commitment will last only so long as the rewards continue or until a better offer comes along. Moral or spontaneous-expressive commitment, in contrast, involves a longer time frame and involves a degree of dedication beyond extrinsic organizational rewards:

> On the one hand, members will remain highly committed to the work system if, and only if, it satisfies the work criteria that is important to them—involvement in decision making, equity, dignity, security, job challenge, autonomy, and so on. On the other hand, the management planners of the work system will continue many of the features of the structure found to be attractive if, and only if, the work system proves to be highly effective in economic terms. (Walton, 1980)

The important work criteria pointed out by Walton are related to each other. The primary focus of these criteria concern meaningful work, trust and participation. To the extent that these processes exist in an organization, they reflect the underlying management philosophy regarding employee motivations and the values. It is, in essence, redistributing authority, responsibility and control downwards to include employees who traditionally exercised little or no control over their work. In its simplest form, such involvement is concerned with getting an employee to express what he or she thinks to the immediate supervisor.

Although it would be tempting to attribute attempts to increase employee control to altruistic motivations on the part of the organization and its management, in reality, such programs exist because of their economic success. As pointed out above by Walton, the bottom-line criterion for management is economic return on their human investment. Employee participation, in other words, becomes valued and promoted to the extent that the organization benefits.

2. *The structures of employee participation*

From a traditional managerial point of view, such programs as job enrichment, quality circles and participatory decision-making pose an inherent threat. By promoting employee involvement in areas that are traditionally ascribed to management, the supervisor tends to see himself as losing authority, control and influence over his employees. Taken to the extreme, the manager sees increasing employee control over work activities as a replacement of his position. This attitude, commonly referred to as a 'zero sum' approach, is based on the belief that the amount of power and influence existing within an organizational unit is finite. Increasing the influence or control of one person or group automatically results in a comparable loss in control for another group. The alternative position, taken by many behavioral scientists, is that the increase in influence by one party does not automatically result in a reduction for the other party. As power is shared, the total amount of influence increases. The paradox for managers is that by allowing more employee participation in decision making, the more total control is exerted over the organizational unit.

While this phenomenon does frequently occur, the typical traditional manager is hard pressed to accept it as an organizational reality. For the manager, it is easier to focus on the loss of power than on the potential benefits of that redistribution. For example, if a manager delegates the responsibility for determining lunch schedules to the group instead of assigning them himself, the manager may typically believe that he or she has lost authority within that area. Without hierarchical control over the situation, employees will abuse the situation. The result will be an increase in problems for the manager. If, in contrast, the manager defines the criteria for the group (for example, have at least two people available to answer incoming phone calls) the problems of setting up and monitoring lunch schedules ceases to exist for the manager. He no longer has to be concerned with satisfying individual preferences, schedule changes, or being seen as the 'bad guy' when individual needs are not met. The manager no longer needs to monitor employee adherence to time spent at lunch since the employees waiting for their co-workers to return will provide a much more effective source of control. For their part, the employees will feel trusted and will exert peer pressure on those who abuse the freedom. They will feel that they are being treated as adults. In such situations as this, it is clear that the total amount of control has, indeed, increased to the benefit of all those concerned.

The rules for this type of participation are straightforward:

1. The involvement and participation must be in an area or topic in which the employees will stand to benefit. The issue to be addressed is whether the manager or the employees should control the situation.

2. The manager continues to exert overall control by clearly specifying reasonable performance standards and expectations. If the employees disagree with the manager with respect to the expectations (e.g., 'Why do we need two people to handle incoming calls during lunch, one person is sufficient'), two courses of action are appropriate. Either discuss the issue until an acceptable compromise is achieved or conduct an *ad hoc* experiment to test the validity of the employees' and manager's position.

3. Allow the group members to monitor performance and adherence of other members to the standard. The manager's role is to ensure that the standards are met, not how they are met.

4. If problems arise among the group members in achieving the mutually agreed upon standards, the manager should not reassert control but rather act as a resource to the group in helping them solve their problem.

5. The involvement and participation of employees in decision making should begin with relatively simple and straightforward issues that can be dealt with in a logical and objective fashion. As employees gain experience in decision making and the consequences of their new responsibilities, the issues that they are asked to address can be broadened to those of a more subjective and complex nature. As the experience of employees with participation increases, their effectiveness in terms of such things as monitoring themselves, time needed to make a decision, and dealing with group conflict tends to improve.

6. Nothing works perfectly all the time. Eventually, the system of participation will break down or be abused. These failures should not be used as an excuse to eliminate the entire program. At best, when these situations arise, the manager and employees should approach the problem as an opportunity to adapt and modify the involvement/participation program.

3. *The degree of participation*

Involvement and participation in organizational decision making is not a unitary concept. There are, of course, degrees of participation. Hirschheim (1985) identifies three broad categories:

1. *Consultative:* The participants provide input into the process but the decisions are ultimately made by another person or group.
2. *Democratic:* All participants have an equal voice in the decision making but the implementation of the decision is left to another person or group.
3. *Responsible:* All participants have an equal vote in the decision making and, in addition, are responsible for its implementation.

The type and level of participation that is appropriate for an organization depends upon many factors, including the past history of participation, the nature of the task, the maturity of the employees, the time frame involved, the consequences of errors, the amount of knowledge available, and the extent to which employee support is necessary for activities to be successful.

4. *The process of participation: why it works*

Employee participation, in its various organizational manifestations, has often been promoted as an approach for increasing the effective contribution of individuals and groups to the organization. Despite the evidence of its effectiveness from numerous case studies, experimental investigations and field studies, little attention has been devoted to identifying the underlying psychological and sociological processes which influence the success or failure of participatory programs.

Human motivation, at its most basic level, is based upon individual physiological and psychological needs. Exactly what these needs are, their nature, number, strength, measurement, etc. have been the subject of endless debates among theoreticians and researchers. As Maslow (1970) and many others have pointed out, motivated behavior occurs to the extent that needs are not fulfilled. As one set of needs is substantially fulfilled, another set arises.

For the most part, contemporary Western society provides for the opportunity to fulfill such basic physiological needs as food, health, and safety. Fulfilment of such basic physiological needs preceeds the fulfilment of psychological needs. Physiological needs, once fulfilled, are replaced by psychological needs. Physiological needs, for the vast majority of employees, are fulfilled by means of income provided by work. It is the fulfilment of psychological and social needs that provide the focus of motivation among the majority of organizational employees. Money, as a motivator, takes on a symbolic (i.e., psychological) value once basic physiological needs have been met.

Employee participation in organizational activities clearly falls within the definition of fulfilling a psychological need. The psychological dynamics that affect and are affected by employee participation can best be understood at two levels, behavioral outcomes and psychological reactions.

At the behavioral level, the participation of an individual in an organizational process or event typically results in a higher quality outcome than would otherwise have resulted. There are several reasons for this. First, having a number of people involved in a decision allows for a variety of opinions. These different viewpoints provide a method for reality checking, identification of faulty assumptions, and correction of errors that is unavailable to a single individual in making a decision. The more people actively involved, the greater the amount of information available. The more accurate information brought to bear on a decision, the better the decision or outcome. Increasing participation of individuals, of course, increases the time required to reach a decision. Decisions or acts that require immediate resolution are less amenable to employee participation than those which have a longer lead time.

Another common behavioral result of employee participation has been termed 'synergism'. This refers to the interaction of several people in a common task. The intellectual interaction tends to result in more ideas and insights being developed than would have occurred had the same individuals worked separately. The reason for this is simple. When people listen to other's comments in a group situation, it spurs their own thinking and insight into areas thay may not have occurred to them individually.

For the manager, the behavioral outcomes of participation are less direct and obvious, hence the reluctance of many managers to promote such activities. The benefits, however, are just as real. To restate some of the points made previously, allowing employees to participate and control aspects of their work frees the manager from many of the non-productive, tedious and time-consuming activities that occupy much of his or her time. It also increases overall group productivity as a consequence of the more effective and thoughtful decision making process. The manager, as a result, typically finds that he or she has more time to devote to issues of a broader nature with a longer term focus. The initial reaction of many managers, 'What will my job be if my employees have a major role in deciding and carrying out the work?', quickly becomes irrelevant. Most managers quickly adapt from their role as an overseer, rewarder and punisher to their changed role as a facilitator and resource for the group.

One frequently overlooked advantage of participation from a managerial point of view is that it provides an opportunity for the manager to observe the leadership potential of employees. For the future manager, participation in employee decision making may well act as a crucible for developing interpersonal skills. Such experiences allow the individuals to get peer and supervisory feedback without the onus of a performance appraisal. In this sense, participation acts as an informal personnel selection evaluation and training activity.

At the psychological level, there is little doubt that participation affects both individual motivation and involvement in a positive way. This phenomenon is the result of three, interrelated psychological processes.

First, the involvement of employees promotes a cathartic venting of individual and group opinions, attitudes and thoughts. Expressing their 'two-cents worth' relieves a great deal of pent-up hostility which might otherwise continue to be suppressed. Such suppression of strongly held opinions results in such passive and destructive outcomes as psychological withdrawal from work (apathy), low motivation (doing the minimum required, clock watching) and passive aggressive behavior (rumor spreading, demeaning others). From a behavioral viewpoint, these types of reaction frequently become evident in turnover rates, tardiness, union grievances and needless quarrelling over trivial points.

Second, when individuals take part in decision making by suggesting ideas, making recommendations, planning and participating in implementation activities, they become involved in the success of the activity. The more input and effect a person has had in planning and carrying out a project, the more motivated he or she is to see it carried to a successful conclusion. In a very real sense, the opportunity to participate in decision making binds a person's ego with the result of that participation. The activity becomes the psychological property of those who shaped it and not a task to be accomplished because 'they' want it done. In a similar sense, open group discussion and commitment made among peers carries with it an obligation not easily denied. Decisions made together with one's peer group are much more difficult to criticize than those generated by the organizational chain of command.

Third, the process of participation inevitably generates more information than was previously known or shared. It may also result in identifying areas in which information is lacking. In most cases, the participatory process itself helps reduce the ambiguity of a given situation. The majority of people work poorly in highly ambiguous situations. Individuals tend to avoid such events and react negatively when they do occur. In the absence of information, employees feel insecure. To counter the lack of information, employees frequently generate their own information in the form of rumors. It has been our experience that rumor generated information tends to be of the 'worst case' genre. That is, rumors tend to be pessimistic rather than optimistic in nature.

Fourth, since participation in most cases involves mutual sharing and interdependence of group members, it promotes group cohesion and morale. This by-product of participation often acts to reduce group conflict since the employees have a procedure for airing concerns. Mutual task interdependence of group members, in contrast to reliance on supervisory direction, has a positive outcome for building effective work teams.

Lastly, by promoting the participation of individuals and groups in the decision making process, the organization and the manager provide strong evidence to employees that they are considered as a resource, valued by the organization. They are not just a replaceable 'hired hand'. The resulting sense of self worth is expressed in a 'we' attitude on the part of the employee rather than one of 'us' and 'them'.

Employee involvement in decision making does not always work out. The behavioral outcomes and psychological reactions presented here do not always occur. Participation of employees in the decision making process appears to work better in some organizational situations than in others. Such programs are usually more successful when the following conditions exist:

There is a sufficient amount of time for group discussion.
Different people have pieces of the necessary information, nobody has it all.
The issues are complex and creative solutions are appropriate.
Those participating will be affected directly by the quality of the decisions reached.

The solutions require interdependence among group members.
There exists a degree of mutual trust and respect among the members.
The issues are seen by the participants as important.
The organization takes the process of participation and its outcomes seriously. Actions are taken as a result.

One of the responsibilities of the manager is to assess the likelihood of successful participation. A fine balancing act between the exercise of managerial authority and group responsibility needs to be walked. As previously described, there are several levels of employee participation. The level of participation should vary with the conditions of success cited above.

Whatever the format or level of participation, it needs to be restated that participation is not a panacea for all organizational ills or managerial problems. Rather than an abrogation of managerial responsibilities, the intention of employee involvement programs is to promote the attitude of ownership and responsibilities among all employees. By itself, participation is only a part of the total picture since its focus is on how work is to be accomplished. The other half of the picture is concerned with the design of the work itself, job design.

5. *Some organizational programs to improve motivation and satisfaction*

In addition to involving employees in the management of the organization and improving the leadership style of managers, a great variety of organization-wide programs can be implemented which have the potential for increasing human resource effectiveness. These activities are focused on the entire organization rather than the individual manager or a specific work group. The effectiveness of these programs has been proven in many different organizational settings over a number of years. Although the approaches overlap to a degree, they may be categorized in the following manner:

1. Recruiting and selecting appropriate personnel.
2. Providing a comprehensive, ongoing, training and development program.
3. Identifying clear and achievable goals and standards.
4. Providing feedback, reward and recognition systems.
5. Promoting appropriate organizational and managerial values.
6. Identifying and implementing ways to enhance work and work systems.

Within each of these six areas, a number of different initiatives are possible. The ideas presented below should be taken as illustrative examples rather than a comprehensive listing. As was the case in previous topics, what is appropriate for one organization may be very inappropriate for another organization.

Recruiting and selecting appropriate personnel

No organization will become very effective if the type of person it recruits and employs is inappropriate for its mission and market environment. It is a tragic waste of human and organizational resources to hire, train and place people whose skills, education, value system or attitudes are inconsistent with the reality of the organization and what it does. A shy person, for example, rarely makes a successful door-to-door sales representative. On the other hand, people whose education, attitudes, skills and values are consistent with those of the organization need little incentive to perform at a high level.

It is an unfortunate fact that most organizational recruitment and individual employment search practices are based upon half-truths and misrepresentations on both parts. On the one hand, the organization advertises job vacancies in the most glowing terms, hoping to attract the best possible applicants and market a positive image of itself. On the other hand, the individual seeking employment is motivated to put his or her 'best foot forward'. When the reality of what an 'interpersonal communication auditor' does become clear—collects unpaid bills over the telephone—the employee feels that he or she has been lied to. Similarly, an applicant who presents himself as having 'participated in market research studies'—he goes shopping for groceries—has purposely mislead the company. Both the organization and the individual are cheating each other.

Recommendations

1. To avoid these types of situations, develop a recruitment plan (brochures, presentations, interviews) which defines the job, its environment, and career prospects in a realistic fashion, describing both the positive and the negative features. If an applicant finds such a position unappealing, he or she will usually select themselves out of consideration. The potential savings for such a program in training, placement and productivity costs are immense.

2. Initiate a research study to determine the educational, motivational, demographic and work experience background best suited to the positions you are recruiting for. Keep in mind that you are likely to be selecting people not only for their ability to fill a specific vacancy but also for future career advancement.

One useful way to perform such a 'job analysis' is to gather a group of current employees together and present them with the issue, 'What individual skills, knowledges, or abilities help/hinder you in getting your job done?' The Focused Group Interview technique described in Chapter 9 may be appropriately used for this.

3. Initiate a research study to evaluate the most cost-effective method of attracting qualified applicants. Many organizations invest in media advertising out of habit. Perhaps word of mouth referrals from current employees or advertising in community newsletters would be more effective and less costly than radio, newspaper or employment agency referrals.

4. Develop an exit interview procedure to determine reasons for leaving. An additional benefit of following up with those who performed well is as a potential source of re-employment should they become disillusioned with their current job. A profile of those who were fired might give some clues as to who not to hire.

5. Delegate primary responsibility for recruiting, hiring and developing new employees to the supervisor and group members to whom the new employee would be assigned. This builds team cohesiveness as well as maintaining skill levels for all involved.

Provide a comprehensive, ongoing, training and development program

1. Initiate a study to identify effective and ineffective work behaviors. Based upon the results of this study, develop a counseling and training program designed to increase the frequency of effective behaviors.

2. Based upon employee and supervisory experience develop a series of employee performance guidelines and training aids which summarize the thinking and experience of such people in typical work situations.

3. Develop a 'senior employee' buddy system for on-the-job training. Each new employee would be paired with a more experienced one who would be responsible for his or her development.

4. Cross-train employees on a rotational basis in as many functions as permissible. Not only does this break up the monotony of single functions but builds a cadre of experienced personnel for promotion.

5. Develop a continuing professional education program for managers and supervisors incorporating such topics as: performance appraisal, effective feedback, motivational techniques, problem solving, group dynamics, managing stress, etc. One effective topic is a 'Problem of the week' presentation of an issue that has arisen which is especially illuminating or unusual. Group discussion of the issues and approaches can be highly educational and motivating.

6. Institute a monthly newsletter to heighten the awareness of human factors issues and to provide a series of guidelines for the practising manager. Each newsletter should be no more than one page in length and address such topics as:

How to set goals and objectives
How to counsel employees
Career planning
Communication skills
Motivating the marginal employee
Problem-solving techniques
Group morale
Leadership styles
How to discipline an employee
Promoting creativity
Giving and getting feedback
Conflict management skills
When and how to delegate
The older employee
Dealing with job stress and burnout
The workaholic
Crisis management

A useful variation on this newsletter approach is to include a brief real-life case study. A number of alternative solutions to the case, issue or problem are presented and readers asked to vote on which solution would be the most appropriate. In the following issue, reader feedback is given along with a discussion as to why one solution might be preferable to the others. Employees might also be prompted to submit their own case studies.

Identify clear and achievable goals and objectives

1. Hold joint manager/employee meetings on ways to measure and evaluate employee performance. Establish standards for acceptable, unacceptable and outstanding performance for each objective.

2. So far as possible, avoid individual objectives. Group objectives are much more motivating when peer pressure can be applied.

3. Develop a series of non-changeable outcomes for each of the three levels of performance.

4. Design a public feedback procedure for group performance. Group comparisons might be made. So far as possible, feedback should be short-term, objectively verified, numerical and posted in a open area.

5. Goals, whenever possible, should be set daily. The longer the time period for goal accomplishment, the more slack there is in the performance system. Feedback, similarly, should be given daily. Project assignments, like work assignments, should also be kept on a tight schedule and for the same reasons. If appropriate, a large project should be broken down into smaller elements to increase the sense of urgency.

Provide feedback, reward and recognition systems

1. Establish a group incentive system. Allow group members to decide how to distribute incentive pay among members.

2. Establish a peer group counseling and performance feedback system as the first level of help. Managers would be consulted only after the team has certified the problem as requiring managerial attention.

3. Establish a 'bottom-up' feedback system from employees to their managers. Out of fear of reprisals, this would most likely be an anonymous system.

4. So far as possible, any awards, recognitions or commendations should be handed out by a high-ranking senior official in a public ceremony.

5. Establish a series of non-financial incentives, keyed to individual and group values. One way to accomplish this would be to have a potpourri of rewards from which the employee or group could choose, save up or trade with each other. The symbolic value of such rewards far exceeds the monetary value. Since any symbolic reward tends to lose its motivating value over repeated presentation, you might consider rotating the rewards. Some examples of incentives are given in Table 12-1.

6. Develop a 'Suggestion of the month' award in which an employee and his/her

Table 12-1. Examples of incentives for employees.

Advancement opportunities.
Special award luncheons for employees and family.
First choice of vacation times.
First choice of lunch shift.
Preferred work assignments.
Assignment as acting supervisor.
Prominent posting of names on the bulletin board.
Opportunity for special training.
Opportunity to act as 'tour guide' for visitors.
Free soft drinks for a week.
Wear a badge 'Number 1' for a week.
Choice of workstation location.
Lunch with the Chief Executive Officer.
Wall plaques.
Opportunity to serve on a special task force.
Certificate of commendation.
Letter of recognition sent to employee's family or local newspaper.

manager become equally responsible for developing the idea, presenting it to higher management and guiding the implementation. Both the employee and the manager would share any resulting rewards or recognitions.

Promoting appropriate organizational and managerial values

1. Adopt a 'sunset law' approach to delegation of authority. On an annual basis, all managers must justify why lower level employees should *not* be involved in various decision making activities.
2. When projects or activities fail to accomplish their stated objectives, hold a discussion focusing on 'lessons learned'. One of the key rules here is to not allow blame to be placed.
3. Establish a joint manager–employee training and development program on any of the topics addressed previously.
4. Establish a cross-functional training program for teams of managers and employees.
5. Hold a biannual open house for employees, families and friends.
6. Establish a regular morning stand-up briefing period to discuss the events of the prior day and to plan for the coming day.
7. Require all supervisors and managers to regularly assume the duties of their subordinates for a day.
8. Conduct an annual employee job satisfaction survey. A summary of the results can be fed back to the employees comparing previous survey results with current outcomes. Recommendations might be solicited for solving the most pressing problems identified in the survey.

Identify and implement ways to enhance work and work systems

1. Based on rotational assignments develop a career ladder approach to promotion.
2. Hold an off-site conference of employees on the topic of work design.
3. Devise a 'sunset law' for all work rules and paperwork procedures in which each would have to be justified to remain operational.
4. Develop local, short-term experiments to study changes in the work flow. Tests that might be conducted could include: redesign of the tasks, test of a new operating procedure, redesign of the workspace layout, installation of new equipment. Be careful of the Hawthorne effect in any of these studies.
5. Promote an annual conference of district and regional managers to exchange ideas, share local initiatives and identify new programs to improve motivation and productivity.
6. Keep work groups and task forces down to a few people. Five to seven people, depending on the task, is ideal. If additional skills are needed, these people should be rotated through the group on a temporary basis. Groups larger than ten people tend to lose their identity and break up into sub-groups.

6. Summary and conclusions

The involvement and commitment of employees and managers to the new (or old) work systems is the critical element in their success or failure. It has been a consistent finding that employee commitment and motivation is a direct function of the extent to which they feel an ownership of the system. When the work, work group or organization is seen as an extension of the individual, he or she will devote an incredible amount of energy and creativity to its success.

There are several approaches to developing this attitude towards the organization. The most common, and least effective, is to 'buy' employee loyalty. This approach will work in the short run or until a higher bidder appears on the scene. A more effective method, although more difficult to implement, revolves around employee participation in decision making, team building and non-monetary rewards.

Depending upon the innovativeness of the organizational culture, there are many organizational activities which can be instituted to enhance employee involvement and commitment to the organization. It is our experience that 'successful', established organizations are loath to try many of the more non-traditional approaches. Newer organizations, or those who are on the brink of a major disaster, are much more responsive in this regard.

Chapter 13
Looking Back: Summary and Recommendations

1. Success factors in retrospect

Hindsight, it is often said, is 20/20. In retrospect, many aspects of the automation process that could have been accomplished in a more effective manner seem obvious. At the time, however, time pressure, legal constraints, unfamiliarity and ambiguity all contributed to resulting relationships among the work, the organization and the employees. The ultimate successes (always a relative term) of the Automated Collection System can be attributed to several factors.

Drawing on the experiences of an established private sector model of an automated collection agency allowed the ACS planners to avoid many of the mistakes and wrong turns that could have been made. Much of the software and hardware used in the ACS had been used for several years in a collection environment prior to its modification for a Federal collection system.

The Collection Office Function was ripe for automation. Those procedures and activities that automation accomplishes most effectively permeated the tasks of the COF: establishing and maintaining standardized records, accounting procedures, issuing standardized forms, repeated telephoning, etc. This fact, coupled with the increasing costs and decreasing availability of the human resources necessary to operate the manual system created an obvious application of automation.

The management of the Internal Revenue Service, from the Commissioner and Assistant Commissioners to the Regional Commissioners, were committed to the concept of automation. The existence of such top-down support created the necessary environment of leadership and support to follow through on the automation. Once approved and started, the effort would succeed.

The majority of employees, from the technicians to the line operators, accepted and supported the need for automation. In essence, the employees believed in the system. In many situations, this belief required an incredible dedication of time and effort far exceeding what the organization could demand. Ultimately, despite the factors previously cited, it was the first-line managers, employees and technicians who made the system work.

It is fortunate that this unique combination of factors was, for the most part, coordinated as to time and resources. The success of the Automated Collection System has provided a working model for many other agencies, both for the United States government, state agencies, and other national governments.

Certainly, within the IRS itself, the effects of the Automated Collection System were widespread. Several other major automation efforts, in progress or planned for the near future, will revolutionize the Service. Foremost among these efforts are:

The Automated Examination System, which is equipping revenue agents with portable microcomputers and custom-designed software to perform more accurate, systematic and speedier on-site tax audits. Revenue agents now have the ability to carry with them to the taxpayer's location all the information and forms necessary to conduct an on-site examination.

The Field Automated Information Storage and Retrieval (FAISR) program, which has constructed the hardware and software for optical disk storage of common tax forms. This process will, eventually, eliminate the need for manual entry of taxpayer submitted data. An anticipated future application of this system will be electronic data recovery and analysis of numerical information.

2. *Lessons learned in retrospect*

Many of the factors affecting the implementation of automated systems in the IRS were, again with hindsight, perfectly predictable had we had the time and resources to cull the existing literature and other information resources available to us. Several *post hoc* insights were unanticipated. In planning for the continuing automation of the IRS, we have attempted to address these factors and anticipate their consequences. To the degree that these 'lessons learned' are applicable to other organizations, they are presented here.

Lesson 1. Plan for the unplanned

However much time and effort goes into the planning process, you can never anticipate everything that will arise. There exists a danger of overplanning and overcontrolling the automation with the result that unplanned opportunities to enhance the system will be missed or will have to be ignored. A certain amount of ambiguity is inherent in any new system. The trick is to allow enough flexibility in the planning to take advantage of unanticipated 'targets of opportunity'.

The key to avoiding these situations is to devise flexible plans, general guidelines and back-up plans for critical aspects of the system. It appears to be a common finding that the more you depend on a specific event or outcome, the lower the likelihood that it will occur.

Lesson 2. Talk to each other

In implementing a complex automation system, the inherent ambiguities mentioned in Lesson 1 often result in *ad hoc*, on-the-spot solutions, since specific guidelines and rules cannot be developed for every situation that might arise. Temporary solutions, given tight time demands and pressure for productivity, often have a way of becoming permanent ways of doing things even if they are not the best solution. Once established, these solutions tend to assume an authority of their own.

Communication among all elements involved in the system implementation reduces the inherent system ambiguity by sharing a number of unique experiences. Communicating common experiences and solutions has the effect of standardizing, if not optimizing, a given approach to an issue. The process of communicating among system elements cannot be left to chance. Three factors affect the success of the communication: (1) it must be institutionalized with cross-functional participation required of all parties, even those who are only indirectly affected; (2) the frequency of communication must be high, especially during the early stages of implementation; and (3) the sharing and application of new information must be rewarded. To do otherwise creates an atmosphere of hoarding and discrediting new information because 'it wasn't invented here'.

Lesson 3. Ripples have a way of becoming tidal waves

The implementation of any major system requires modification. All too frequently, what appear to be minor modifications in the system have a ripple effect on other, interdependent, components of the system. While the notions underlying systems theory are widespread, their applications are not. It takes time and effort to think through the consequences that changing one part of the system might have on other parts. In the pressure for implementation, the focus is usually on the immediate consequences. Involving the system's users, a comprehensive planning model and appropriating enough time for its use will cause immediate delays but will reduce the need for large-scale retrofitting and backtracking.

Lesson 4. The user knows best

Although it is temptingly easy to ignore the rank and file employees in the design and implementation of the new system, they are almost always the best source about the aspects that work and don't work. Equally important, they are an invaluable source of information and ideas of how to make the system work better. Too frequently, the people who actually have to work the system are ignored and treated as a replaceable resource. The participation of employees should be encouraged and rewarded. In addition, acting as an information resource, participation will improve the motivation and morale of those who will bear the brunt of decisions made at higher levels. In essence, participation promotes a 'we' attitude rather than an 'us and them' feeling.

Lesson 5. Barnfires are easy, controlled explosions are not

Most major innovations create an initial surge in motivation and morale. The implementation of automation creates high expectancies, willingness to sacrifice, and high levels of dedication among employees. In fact a common phenomenon among employees during the implementation and de-bugging stage is employee 'burnout'.

The critical dimension of the successful automation effort is to maintain a high level of motivation and dedication over time. Satisfaction and motivation levels tend to stabilize quickly after implementation. What was novel becomes routine. The most difficult problems to deal with are the long-term maintenance needs, not the short-term implementation concerns. Unfortunately, the long-term focus is usually ignored or put on the back burner given the near-term problems of getting the system up and running.

Lesson 6. Try it before you buy it

Given the amount of resources that are to be dedicated to implementing automation, it would seem reasonable to conduct a pretest or pilot test of systems operation before the decision is made to implement. The pilot test should be for the sole purpose of experimenting with alternative ways of designing the final system. The emphasis on getting the job done while pilot testing is inappropriate and often limits the freedom to innovate. In practice, the pilot test should reward innovative approaches and promote an attitude of learning from the mistakes that should be made.

Lesson 7. Work for people

The fact that people are incredibly flexible, both mentally and physically, has often been taken advantage of by system planners. Except in extreme cases, people will usually adapt to the situational demands, at least for a short interval. And then many will leave the system. Poor ergonomics, limited career opportunities or pay increases, unremitting pressure for bottom-line results and ineffective management will take their toll. Those who are able to will leave the system for better prospects.

In addition to designing equipment, workspaces and software around human capabilities, it is critical to incorporate the human element in the design of the work itself. For tasks that are inherently boring, it is often necessary to plan for a work team approach—job sharing, job rotation or job enrichment programs. For jobs with limited flexibility, design a clear career ladder with measurable performance criteria and a short time frame.

Identify and recruit for the type of employees that are willing and competent to work in an automated environment. Realistic recruiting practices, while likely to reduce the actual number who apply and accept the vacancy, will pay dividends in lowered attrition after a costly training program. Broadly speaking, two types of employee are attracted to the automated environment: those for whom the position is viewed as a temporary means to an end (i.e., a 'better' job) and those for whom the job is an end in itself. For the first type, a career ladder needs to be clearly defined. They are willing to put up with a position which demands little of their skills for a brief period before moving on. For the second type, the emphasis needs to be placed on rewards (social, financial, institutional) that are available for high performance in their current position. Any viable organization needs both types of employees since they provide for both stability and growth.

Lesson 8. Change managerial behavior before you attempt to change attitudes

Automated work environments require a different set of managerial skills than a manual environment. There is, however, a strong tendency to apply the managerial styles that were perceived to be successful in the prior environment to the new environment. The typical organizational response to this is, 'We need to train managers in human relations skills'.

Organizations put a great deal of money, time and effort into training managers in a variety of topics which go under the general title of human relations skills. The intention of these programs is to provide managers or managers-to-be with insights into the nature of human behavior and motivation. However well intended, planned or presented, it is the author's perception that, except in rare instances, these attempts are generally ineffective.

The assumption underlying human relation training for managers is that the skills developed in a classroom, training laboratory or university setting will transfer to the workplace. The model for this approach is that of technical skills training. Within the educational setting itself, significant changes in attitude and values may occur. Where the system fails is in the transfer process to the actual work environment. This failure of the transfer of training process may be attributed to two fundamental factors.

First, by their very nature, instructional environments tend to emphasize deductive learning. General principles or theories are espoused. It is on the shoulders of the participants to learn how to apply these principles in specific and unique situations. Lacking a great deal of guidance and support, managers tend to fall back on their traditional ways of managing. People appear to find it difficult to operationalize general, often ambiguous, principles.

Second, whatever attitudinal changes occur in the somewhat cloistered environment of the training situation are generally not supported or reinforced in the work environment. Attitudinal changes, in the face of the often harsh realities and demands of the workplace, tend to be short-lived. Permanent shifts in attitudes, values and beliefs regarding the role and nature of employees are unlikely to occur within the relatively short time span of a training session. At best, one would expect a suspension of prior assumptions and a willingness to experiment with a different approach. The ability to shed strongly held beliefs overnight almost never occurs and should be looked upon with a doubting eye when it does. Usually, authentic changes require time and a series of positive work experiences to support them. Few human relations training programs are planned to provide such long-term support.

It is not uncommon for organizations to react to the frequent long-term failure of human relations training programs by requiring more training or continually searching for 'better' programs. Doing the wrong thing harder will not make it right.

An alternative approach to human relations training might be based on an inductive learning model. To accomplish this it is necessary to reverse the traditional order in which training events occur. In its very broadest outline, this approach might be somewhat like the following:

Step 1. Identify managerial behaviors which are appropriate for the work situation and consistent with human relations principles.
Step 2. Establish a formal organizational system to ensure managerial compliance with the required bahaviors.
Step 3. Monitor managerial behavior and reward compliance.
Step 4. After compliance becomes routinely established, hold an informal managerial meeting to discuss the human relations concept underlying Step 1.
Step 5. Promote new applications for the principles identified in Step 4.

In sum, successful behavioral change often results in attitude change while the opposite is not usually true.

Lesson 9. Plan and budget for the unforseen

Even with the best upfront planning, with the utmost input from all specialized fields that need to be involved, and with past experience as a guide, the need to maintain a budget 'cushion' for retrofitting will exist. A balance between identifying the minimum resources required to get the job done and a contingency budget for unforeseen events needs to be decided. The automation effort that strictly allocates every resource will, almost invariably, find itself hamstrung.

Lesson 10. Success is in the long run

The novelty of new work technologies combined with the increased range of discretion and personal involvement all combine to create a high level of motivation during the implementation period. This high level of dedication is a temporary phenomenon. If the effect of automation is to simplify the tasks, reduce individual judgment or reduce work cycle time, the period of time until motivation reaches a plateau and starts to drop can be very short. The real test of success is the ability to maintain high levels of commitment over the equipment life cycle. The more routinized the tasks, the more motivational stimuli must come from such work enviroment aspects as co-worker pressure, organizational incentives, feedback systems and leadership styles.

The intent of a case study such as this is to convey the conditions, successes and failures which resulted from some series of events. In the broadest sense, these ten 'lesson learned' summarize the events and conclusions described throughout the rest of the book. As evidence accumulates from different situations and times, hopefully the lessons learned will be transferable. It has been our experience that including human factors considerations *a priori* is almost always more effective than dealing with them in a *post hoc* fashion.

Appendix 1
Recommendations by the Human Resources Technology Staff

Introduction

Reacting to a concern that human factors were not being given appropriate attention as the Internal Revenue Service moves into automation, a working group was established to make recommendations on steps to be taken to address this issue.

The group has identified and will be focusing on five basic approaches to include human factors considerations in an automated environment. These five approaches will be introduced and developed during the Automated Collection System (ACS) pilots in four districts. Because of the advanced status of the ACS pilots and because of the vendor competition, such activities will be limited. Techniques developed during the pilots will be refined for use in other Service automation efforts.

The five approaches discussed in this paper are as follows:

Institute an Employee/Manager feedback system
Increase Employee/Manager interpersonal contact
Introduce new Orientation and Training modules relevant to preparing for technological change
Design innovative working arrangements to humanize work
Seek Labor/Management cooperation through early union involvement

The attached outline encompasses a series of proposed actions to incorporate human resources factors into automated systems planning and implementation. The recommendations were designed to be independent of each other. That is, the potential exists to delete one or more recommendations without seriously affecting the implementation of other recommendations. The one exception would be that actions which require the acquisition of new knowledge and skills would require training.

The outline itself is comprised primarily of two broad topic areas: The first area describes Phase I recommendations primarily concerned with the ACS pilots. Phase II recommendations will be initiated during Phase I and describe long term experiments extending beyond ACS. Phase I and Phase II of the outline focus on specific areas of concern to human resources and present recommendations to address these issues. In addition, Sections III and IV list some anticipated results of our proposals and outline Labor/Management Relations Implications.

Human Resources Technology Staff recommendations

I. Initial scope of actions to be implemented beginning now and continuing through ACS pilot

Given Assumptions:

No additional National Office personnel evaluation teams at pilot sites as a result of this proposal.
No significant work interruptions during duty hours.

Employee/manager feedback system

Employee feedback systems are designed to accomplish four broad objectives:

Provide a mechanism for identifying problems arising out of the use or anticipated use of automated equipment and potential resolutions to these problems.

Provide 'base line' data for assessing and evaluating the impact of automated equipment on organizational functions within a wide range of variables.

Provide an opportunity for users to suggest new applications and approaches for enhanced organizational effectiveness, e.g., tasks, procedures, group processes, structure.

To detect further training needs, refinements, etc.

The following systems would utilize a micro-computer to input, FTS lines for transmission (estimated to be up to one hour per day), and a receiving micro-computer in the National Office during the ACS pilot period. Once the vendor is selected, use of ACS equipment should be explored.

The National Office will purchase or contract for the purchasing of the computers for the ACS districts in addition to the one to be located in the Human Resources Technology Group in Washington. National Office Collection and the National Office ACS Project Team will each have a micro-computer linked to this system. Collection and ACS will be utilizing the system to transmit data related to their respective evaluation processes.

1. Electronic Employee/Manager Feedback

Employees and managers would be asked to record problems or suggestions for improvements on a specially designed card. These cards would then be collected and fed to the National Office work group via a computer link. The group would use these employee reactions to identify: (1) changes in employee attitudes toward work, the environment, supervision, the social atmosphere, etc.; (2) procedural, mechanical or systemic equipment problems; and (3) additional training needs.

In addition to the identification of immediate and potential problems and opportunities for improvement, results could be used in the design of long-range planning for future automation projects.

This concept could also allow for employee feeling of involvement in the test process outside of equipment testing. It may well also offset to some extent some of the negative aspects of a production oriented system.

In order to protect the vendor evaluation process, it may be necessary to pool information from all call sites in order to erase site identification. If this concern could be overcome, there is some advantage to evaluating data with site identification in the interest of overall test validation.

Such a system might also be used by the ACS System Acceptability Test (SAT)/Evaluation Team as a direct information feed to the National Office Automated Collection System organization.

It is also recommended that a system be explored to utilize the vendor's computers to provide periodic feedback to employees on the results of their contacts (such as dollar amounts of payments made).

2. EMPLOYEE QUESTIONNAIRE

An employee questionnaire would be designed to gather data on changing employee attitudes toward all aspects of their work—environment, comfort, supervision received, space, social atmosphere, etc. The questionnaire would be administered on the first day of training, at least once during the pilot, and following the pilot.

The content areas of the questionnaire would include the following:

Organizational communication patterns
Organizational leadership styles
Job satisfaction
Group processes
Feedback systems
Career orientation
Sociodemographic factors
Organizational climate
Ergonomic concerns

The questionnaire is designed to be self-administered in about 15 to 20 minutes. Responses will be both closed and open-ended. The results of this questionnaire will be computer and content analyzed for patterns of responses.

3. EXIT INTERVIEWS

During the course of the pilot project some attrition of employees is expected. While people leave a job for a variety of reasons, those reasons which arise from pilot working conditions need to be documented and evaluated.

To supplement the prescribed exit interview process, we propose to develop an interview guide. This interview guide will be focused on issues pertinent to the pilot program. More specifically, we will attempt to have the interviewer gather narrative information on ergonomic, social and administrative factors which contributed to the decision of the employee or manager to withdraw from the ACS pilot program.

Such information will be collected as an adjunct to the normal exit procedures. An interview guide will be developed by the Human Resources Technology group and distributed to appropriate staff at each test site for administration as attrition occurs. Completed forms will be forwarded to the National Office Staff following each interview.

Employee/manager interpersonal contact

Decreased personal contact due to the new automated system dictates a need for structuring alternatives for increased interpersonal dealings. Studies of computer-mediated work and offices with information technology systems indicate that people can feel isolated in an impersonal situation. They may also feel they have lost the support and assistance they previously received from communication with peers and supervisors. These feelings can hamper the productivity goals of the organization if they result in reduced job satisfaction, job commitment and appropriate risk taking.

We recommend that the following actions be considered by local managers:

1. Managers review their current practices for managerial staff meetings and employee–supervisor individual and group meetings, then determine any obvious improvements in local communication and problem-solving they feel are necessary under the existing work environment. Given the increased data available under the ACS system, they can project any improvements they wish to implement under the new system.
2. Recognizing that a 43% overhead figure has been built into ACS staffing we recommend that managerial, employee and employee/manager team meetings should be held during duty hours. Meetings could be held periodically or weekly as needed. Length would be determined by agenda, but normally would be no longer than one hour.
3. Purposes of the meetings would be defined by local need. However, in addition to traditional information sharing meetings, options appropriate for the new work environment include:

   Problem solving meetings in which employees are notified in advance to identify problems and solutions and then meet as a group with their supervisor to resolve them.

   Coverage of issues above and beyond procedural and equipment issues such as communication, interpersonal dealings with taxpayers and others, and opportunities for improvement in the work environment.

   Team building, goal setting and group feedback sessions during which results of the unit's work are discussed and accomplishments are recognized timely.

Limits would be set as to the scope of recommendations or changes the group would implement to ensure a fair test during ACS vendor competition. If such meetings represent a change in the local site's normal practices we recommend they be initiated first at managerial levels and subsequently at the employee level. This would provide managers with increased familiarity with the processes involved and greater confidence and skill in facilitating employee meetings.

Techniques for conducting and participating in these meetings are recommended for managerial and employee training.

Feedback reporting systems

Information will be obtained throughout ACS training and the pilots. Information sources will include the Employee/Manager Feedback System, the questionnaires, ACS group meetings, exit interviews, and perhaps others not yet defined. Data obtained will be periodically analyzed and compiled into reports for submission to Human Resources management in the National Office, National Office Collection management, ACS National Office Project managers, pilot site regional officials, and others as required. This work group will serve as an information resource throughout the period covered by the implementation of proposals adopted.

Orientation for ACS managers and employees

Changing from the current work environment to a fully automated work environment presents a significant change to employees and managers who will work with ACS and other automated systems. Organizations and consultants who are familiar with similar changes in work environments have identified the need for adequate information and preparation of the users to manage this change successfully.

We recommend a several-step approach to orient managers and employees to this new work environment and ACS. The steps are:

1. BRIEFINGS CONDUCTED BY IRS MANAGEMENT

Between now and March 21 (the earliest date ACS technical training will begin) we suggest that all new ACS employees and managers who have not already received a briefing on ACS be scheduled to attend briefings, we also recommend that current employees receive appropriate updates.

Timing and coverage would need to be coordinated with similar actions already scheduled for training. Content could include:

How ACS was designed to support the IRS mission.

ACS organizational structure and specific goals.

Projected advantages of the automation for the organization, managers and employees.

Showing ACS slide-tape presentation (for those new to IRS).

An update on the current status and action plans for equipment installation and training schedules at each site.

National NTEU representatives would be informed in advance of these briefings. Appropriate local NTEU representatives would be invited to attend local briefings.

Managers designated to conduct the briefings would depend on the number and work assignments of the employees. Where practical, we recommend that the call site Director and the ACS Project Coordinator jointly conduct the briefings.

2. ON-SITE ORIENTATION TO EXISTING IRS MICRO-COMPUTER OPERATIONS AND ACD SYSTEMS

Those sites which now have micro-computer equipment in operation (for example, Compustar) would provide orientations to employees and managers either on-the-clock or during lunch and learn sessions.

Information would be shared on the capacity and application of the existing equipment.

Where feasible, hands-on experience to enter and retrieve data in the system could be provided for small groups.

Those sites which have Automated Call Distribution (Collins-Rockwell) systems would provide orientations.

3. VISITATIONS TO LOCAL AUTOMATED WORK SITES IN PRIVATE INDUSTRY

We recommend that a representative number of key ACS personnel visit an automated system in private industry or government. Visits would be designed to increase familiarity with an extensively automated office setting and would take place prior to the start of training.

Selection of the site and arrangements for visits would be coordinated by local call-sites. Factors to consider in selecting a site might include the following:

Successful implementation of office automation.

Similarity of the work and automation equipment to ACS (for example collection agencies, banks, insurance companies).

Organizational climate and attention paid to both human factors and equipment technology.

Extent of experience with automation beyond word-processors.

Proximity to the call site and cost of any travel for the visitation.

Availability and willingness of the host site to provide an introduction or briefing to IRS visitors.

We recommend that National Office ACS and Human Resources functions identify any known sites close to the call-site districts.

4. VOLUNTARY AUTOMATION ORIENTATIONS

An additional alternative would be for ACS sites to provide information to employees and managers about opportunities for individuals to increase their familiarity with automation. Suggested actions would take place during lunch or break periods and could include:

Bulletin board notice of micro-computer demonstrations sponsored by vendors and computer groups in the community.

Training and Development Branch Library pamphlets, books and displays on micro-computers.

Identification of scheduled public television shows which periodically cover advances and applications of computer technology.

Purchase and showing of existing video-tapes (for example the BBC Computer Program or the Boeing series) which introduce computer applications for office and outside-the-office use. Information on current programs could be identified by National Office and Regional Training and Development functions who have reviewed, purchased or developed ADP and Information Technology training and orientation modules during the past year.

Training

The ACS pilot provides a vehicle to increase involvement of employees and managers in identifying problems, solutions and innovative opportunities with the new automated system. We recommend that training be offered to encourage innovation, enhanced productivity, and participation of the users during the pilot. Based on experience with similar approaches in other organizations, potential benefits include:

Enhanced knowledge and skill in problem resolution.

A work climate that goes beyond implementation and includes focus on improvements and innovation.

Improved use of resources in having a problem-solving function performed by more people who have direct experience and different kinds of knowledges and experience than evaluation teams and higher management alone.

Increased feelings of control over the work environment to compensate for the feelings of apparent loss of control frequently encountered by groups working with new technology.

Increased sense of ownership in the goals, problems and solutions of the organization.

Greater commitment to implementation of solutions.

Enhanced confidence in performing the work due to contributions individuals can make to use technology as a tool.

Reduced non-productive frustration and anxiety that might accompany introduction of technology without these human factors measures.

1. TRAINING FOR BOTH MANAGERS AND EMPLOYEES

(*a*) Methods for problem finding and problem solving. This would include:

Observation skills—what to look for in terms of problems and opportunities in the new operation, new equipment, new work assignments and new organizational structure and transactions.

Techniques for diagnosing problems in the flow of work and communication among people and work units.

Techniques for diagnosing problems in tasks and programs.

How to facilitate the problem-solving and decision-making process in working with groups (employees, peers, bosses) to ensure implementation and ownership in problem resolution and innovations.

(*b*) How to use the electronic Employee/Manager Feedback System.

(*c*) Managing Change and Managing Stress (topics have already been scheduled for development as part of the ACS managers' training).

2. FOR MANAGERS ONLY

(*a*) Techniques to facilitate interpersonal communications in view of the new automated environment and demand for continuous computer operation.

(*b*) Job redesign and organizational restructuring awareness to identify suggestions for implementation *after* the pilot.

Ongoing communication and support actions

Communicate top management support for increased involvement of employees and managers in problem solving and innovation prior and during the pilot.

For example, a letter could be sent from the Commissioner to employees at all four sites, and a letter sent from each call site Director highlighting ACS as a front runner in IRS efforts to incorporate human resources factors and new automation technology.

Information bulletins (like the ACS Newsletter) could highlight new information and problem solving successes yet be designed to protect vendor competition.

Specific written and oral communications to managers on productivity goals could include reference to human resources innovations as well as traditional goals (e.g., expedited implementation of sound employee suggestions, productive use of electronic feedback system).

Communicate tolerance for local experimentation to resolve problems (within limitations needed for vendor test) at staff meetings.

Recognize and praise innovations in task and work group management.

Study design

The actions proposed in this paper are designed to promote human resources considerations in the implementation of new office technologies. To evaluate such actions, we propose to compare and contrast several aspects of our data bases. Primary analyses will include the following:

Questionnaire comparisons: pre-training responses compared to mid-study responses and to end-of-pilot responses.

Electronic or written employee/manager feedback: month to month evaluation and summary by test site.

Employee–manager interpersonal contact: Bi-monthly comparisons by test site followed by overall summary at the end of pilot.

Orientation and training: Evaluation during and following classroom and on-the-job training followed by summary at the end of pilot.

Exit interviews: Tri-monthly comparisons by test site followed by overall summary at the end of pilot.

The use of data comparison groups allows for an additional comparison of results by providing a base line against which changes can be compared. The agreed-upon approach to establishing control groups is as follows:

The proposed systems will be placed in each of the four test sites. A comparison will be accomplished by administering only the questionnaire in the non-ACS districts (Boston and Denver) and comparing results. It would thus be possible to distinguish between problems attributable to ACS or those which are not.

II. Anticipated results of proposals

Proposed actions

1. Employee feedback
2. Increased employee/manager interpersonal contact
3. Orientation and training
4. Actions taken to humanize work
5. Labor/management co-operation

Anticipated results

Employees and managers understand the change

They understand their new roles and responsibilities

They understand the philosophy behind the technology

They accept the change

There is no or minimal resistance

Conflict is resolved or negotiated

They believe in the change (that it is necessary and beneficial)

They feel comfortable with the change

They feel the organization is committed to the change

They are comfortable with and committed to their new roles and responsibilities

Enhanced productivity

III. Labor/management relations implications

It is recognized that this project will have significant labor/management relations impact. The gathering of direct feedback from employees, the utilization of employee/management problem solving sessions, any surveys, questionnaires, etc., are all items which require that due deference be paid to the institutional rights of the employee's union. Indeed, many of the proposals may require actual negotiations with NTEU.

RECOMMENDATION

To have NTEU receive periodic briefings similar to the Multi-Functional Workforce briefings.

With this approach, the union will be aware of upcoming events so that they can surface their problems or concerns early in the process. They can also be made aware of when to expect formal notice of changes, innovations, etc., so that both parties can avoid the unnecessary litigation which often results from last minute notice.

We also recommend that an initial briefing be scheduled for sometime in January or February, 1983, as soon as the initial decisions are finalized as to the overall direction which this project is to take. If NTEU opts for briefings, such sessions may be scheduled on an as-needed basis, but we envision that it may ultimately be necessary to hold them on a regularly scheduled (e.g., monthly) basis in order to maintain the union's 'comfort level' with the project.

Appendix 2

What's Your Opinion??? Form D

Directions

Your direct involvement with the Automated Collection System over the past several months has provided you with the opportunity to become increasingly familiar with procedures and systems. This questionnaire, like the previous ones, gives you the opportunity to express your opinion regarding various aspects of the Automated Collection System. The results of these surveys will be put together in a report with recommendations for future ACS operations. Your opinion DOES count.

As is the case for the entire study, no individuals will be identified. Only group averages will be reported.

Turn to the following page of the questionnaire and place your 4 digit survey number in the box provided at the top of the page. If you cannot remember your Survey Number and do not have the Participant Card given to you at the beginning of the study, raise your hand. The proctor will give you the envelope containing your survey number.

Space has been provided in several sections of the questionnaire for your written comments regarding your answers. We encourage you to use these comments sections.

Thank you for taking the time and effort in filling out this survey.

Survey Number: ....

I. Background

1. **Current ACS job title:**

Account Representative	...... (1)
Account Investigator	...... (2)
Research Analyst	...... (3)
Clerical	...... (4)
Contact Unit/Section Chief	...... (5)
Research Section Supervisor	...... (6)
Investigation Section Supervisor	...... (7)
Other	...... (8)

2. **Job title prior to ACS:**

Non-IRS (previous job title? ....)	...... (1)
Clerical	...... (2)
T.A.R.	...... (3)
O.C.R.	...... (4)
Revenue Representative	...... (5)
Process Review	...... (6)
T.S.R.	...... (7)
T.S.S.	...... (8)
C.S.B.	...... (9)
Exam Division	...... (10)
C.I.D.	...... (11)
Other (job title? ....)	...... (12)

3. **Current Work Status:**

Full time permanent	...... (1)
Part time permanent	...... (2)
Full time temporary	...... (3)
Part time temporary	...... (4)
Full time seasonal	...... (5)
Part time seasonal	...... (6)
Full time intermittent	...... (7)
Part time intermittent	...... (8)

4. **Current grade (for example, GS 5):** ....

5. **Age:** ....

6. **Sex:**

   .... Male
   .... Female

7. **Which of the following best describes your status as a wage earner?**

Secondary wage earner in a household of 2 people	...... (1)
Secondary wage earner in a household of more than 2	...... (2)
Primary wage earner in a household of 2	...... (3)
Primary wage earner in a household of more than 2	...... (4)
Sole wage earner supporting self	...... (5)
Sole wage earner supporting self and others	...... (6)
Other (specify)	...... (7)

8. How did you first learn of the opportunity for employment with ACS? (check one)

Don't recall	...... (1)
Newspaper ad	...... (2)
Radio ad	...... (3)
Employment Agency	...... (4)
T.V. ad	...... (5)
An IRS employee	...... (6)
Vacancy notice	...... (7)
Other (specify)	...... (8)

9. What first attracted you most to employment with ACS? (check all that apply):

Continued IRS employment	...... (1)
Salary/benefits	...... (2)
Hours of work	...... (3)
Closeness to home	...... (4)
New work technology	...... (5)
New work environment	...... (6)
Nature of the work itself	...... (7)
Short term employment	...... (8)
Learning opportunities	...... (9)
Other (specify)	...... (10)
....	

10. Within the next few years I would expect that: (check one)

I will be working at the same job I now have	...... (1)
I will be promoted within this job series	...... (2)
I will stay in IRS but leave ACS	...... (3)
I will be working at another state/federal job	...... (4)
I will be working in private industry	...... (5)
I will not be working	...... (6)
Other	...... (7)

11. Which factors would be most likely to influence you to leave the ACS? (check all that apply):

salary/benefits	...... (1)
hours of work	...... (2)
transportation problems	...... (3)
nature of the work itself	...... (4)
advancement opportunities	...... (5)
production demands	...... (6)
work environment	...... (7)
no plans to leave	...... (8)
Other (specify)	...... (10)
....	

12. Highest educational level:

Non-high school graduate	...... (1)
High school or GED degree	...... (2)
Some college work	...... (3)
Associate degree	...... (4)
Bachelor's degree	...... (5)
Some graduate work	...... (6)
Graduate degree	...... (7)

13. Total months in IRS:

Less than 2 months	...... (1)
Between 2 and 6 months	...... (2)
Between 7 and 12 months	...... (3)
Between 13 and 18 months	...... (4)
Between 19 and 24 months	...... (5)
Between 25 and 36 months	...... (6)

II. Work Process

In the following series of questions you are asked to indicate how satisfied you are with specific aspects of your work process by placing a check mark under the appropriate column heading. If you have no opinion or the question is not applicable to your job, place a mark in the "No opinion/NA" column.

How satisfied are you with the following aspects of your work: (check once for each aspect)

	No opinion/NA	*Very unsatisifed*	*Unsatisfied*	*Satisfied*	*Very satisfied*
14. The total amount of your workload:					
15. Changes in workload (peaks and valleys):					
16. The speed or pace at which you are expected to work:					
17. The emphasis on productivity:					
18. The amount of challenge in your work:					
19. The amount of interruptions:					
20. The amount of feedback and help you get from your supervisor:					
21. The amount of feedback and help you get from your co-workers:					
22. The ease of "housekeeping" chores (records, filing, paperwork, etc.):					
23. The amount of paperwork your work involves:					
24. The amount of control you have over what you do and when you do it:					
25. The availability of help when you have a question or problem:					
26. The variety of things you do:					
27. IRS classroom training:					
28. Non-IRS training provided by a contractor or vendor:					
29. On-the-job instruction:					
30. Training and reference manuals:					

Comments? ______________________________

____

____

III. Work Environment

The questions in this section attempt to measure your reaction to your work environment. For each item below, indicate how satisfied you are with that aspect by placing a check in the appropriate response column.

	No opinion/NA	*Very unsatisfied*	*Unsatisfied*	*Satisfied*	*Very satisfied*
31. Temperature and humidity levels.					
32. Air ventilation and circulation:					
33. Noise levels (general background). If unsatisfied, please specify source of noise in comments section below:					
34. Noise levels (work station area):					
35. Arrangement of furniture and work areas:					
36. Overhead lighting:					
37. Work station lighting:					
38. Natural lighting (windows):					
39. Orderliness of the room and materials:					
40. Accessibility of office supplies:					
41. Availability of office supplies:					
42. Break areas:					
43. Restroom facilities:					
44. Walkways between work areas:					
45. Use of color and paint in office:					
46. Use of pictures and plants in office:					
47. Ability to "personalize" your work station:					
48. Overall satisfaction with work environment:					

Comments? ______________________________

____

IV. Work Equipment

This section deals with the equipment, materials, and furniture you might use to accomplish your work. For each item below, indicate your degree of satisfaction with that aspect.

	No opinion/NA	Very unsatisifed	Unsatisfied	Satisfied	Very satisfied
Work Station Desk:					
49. Ease of adjustment:					
50. Range of height adjustment:					
51. Size of work surface:					
52. Personal drawer:					
53. Leg space:					
54. Wrist support ease of adjustment:					
55. Wrist support range of adjustment:					
Comments? ______					
Work Station Chair:					
56. Ease of height adjustment:					
57. Ease of back adjustment:					
58. Range of height adjustment:					
59. Range of back adjustment:					
60. Ease of chair movement:					
61. Overall comfort of chair:					
Comments? ______					
Mobile Storage Unit:					
62. Ease of movement:					
63. Personal drawer operation:					
64. File drawer operation:					
65. Amount of record space:					
66. Ease of lock operation:					
67. Degree of security provided by lock:					

	No opinion/NA	Very unsatisifed	Unsatisfied	Satisfied	Very satisfied
Personal Effects Locker:					
68. Adequacy of size:					
69. Ease of lock operation:					
70. Degree of security provided by lock:					
Comments? ______					
Video Display Unit:					
71. Legibility of letters and figures:					
72. Screen glare and reflection:					
73. Screen contrast:					
74. Screen angle and position:					
75. Screen brightness:					
76. Screen color:					
77. Overall ease of CRT use:					
Comments? ______					
Keyboard:					
78. Positioning of keys on keyboard:					
79. Positioning of the keyboard:					
80. "Feel" of quality of keyboard entries:					
81. Overall ease of keyboard use:					
Comments? ______					
Software Program:					
82. Amount of information presented:					
83. The order in which information is presented:					
84. Ease of information entry:					
85. Number of errors made in entering information:					
86. Layout and format of screen information:					
87. Ease of error correction:					

V. General Issues

In the following section are several items which relate to some of the more general aspects of your work, what you do and how you do it. Although each question has a space for comments you may use the reverse side of the page for additional space.

89. Everything considered, I am satisfied with working for the IRS.

No opinion	...... (0)
Strongly disagree	...... (1)
Disagree	...... (2)
Agree	...... (3)
Strongly agree	...... (4)

Comments? ....

90. I am satisfied with doing the work my job requires.

No opinion	...... (0)
Strongly disagree	...... (1)
Disagree	...... (2)
Agree	...... (3)
Strongly agree	...... (4)

Comments? ....

91. I am satisfied with the spirit of teamwork that exists between co-workers.

No opinion	...... (0)
Strongly disagree	...... (1)
Disagree	...... (2)
Agree	...... (3)
Strongly agree	...... (4)

Comments? ....

92. I am satisfied with the support my supervisor provides me.

No opinion	...... (0)
Strongly disagree	...... (1)
Disagree	...... (2)
Agree	...... (3)
Strongly agree	...... (4)

Comments? ....

93. There is a bottleneck in my organization that seriously affects the flow of work.

No opinion	...... (0)
Strongly disagree	...... (1)
Disagree	...... (2)
Agree	...... (3)
Strongly agree	...... (4)

Comments? ....

94. In general, this job is: (check one)

Much worse than I expected	...... (5)
Worse than I expected	...... (4)
About what I expected	...... (3)
Better than I expected	...... (2)
Much better than I expected	...... (1)

Comments? ....

95. About what percent of your work time involves doing things that are not directly related to getting your job done?

...... %

Comments? ....

96. How useful to you is the training you received on "managing stress and change" in your current job situation?

No opinion or NA	...... (1)
Not useful at all	...... (2)
Not very useful	...... (3)
Moderately useful	...... (4)
Very useful	...... (5)

Comments? ....

97. Compared to the "best you could do in this job under the current conditions, how productive would you judge yourself to be? (Circle the appropriate percentage on the scale below)

....

0% 10% 20% 30% 40% 50% 60% 70% 80% 90% 100%

98. Compared to the performance level your supervisor expects of you, how would you rate your performance?

Well below acceptable	...... (1)
Just below acceptable	...... (2)
Acceptable	...... (3)
Just above acceptable	...... (4)
Well above acceptable	...... (5)

99. Compared to where you are now within the next few months would you expect your performance to decrease, stay about the same, or increase?

Performance decrease greatly	...... (1)
Performance decrease slightly	...... (2)
Performance stay the same	...... (3)
Performance increase slightly	...... (4)
Performance increase greatly	...... (5)

100. Which, if any, of the following factors limit your productivity? (check all that apply)

lack of management support	...... (1)
emphasis on productivity	...... (2)
work too tedious or repetitive	...... (3)
hostile taxpayer contacts	...... (4)
lack of taxpayer account information	...... (5)
too much time on terminal	...... (6)
lack of collection or tax law knowledge	...... (7)
slow page times on computer	...... (8)
other (specify)	...... (9)
....	

101. What, if any, additional experiences, training or practice would help you do your job better? (Check all that apply)

On-the-job instruction	...... (1)
Specify:	
....	
Interview techniques	...... (2)
Specify:	
....	
Simulated conversations with taxpayers	...... (3)
Objective performance feedback	...... (4)
Keyboard entry skills	...... (5)
Talking and entering data at the same time	...... (6)

102. What, if any, additional experiences would you recommend for future call site personnel like you as they prepare to come on line?

Orientation visit to a working ACS call site	...... (1)
Orientation visit to a near-by non-IRS automated office	...... (2)

Comments: ____

____

____

103. What, if anything, should be done to improve the system hardware and software?

Problem	Solution
1.	....
2.	....
3.	....
4.	....
5.	....

104. Please rate the following training topics using the following scale:
   0 if the topic is not relevant to your job
   1 if there was too little training on this topic
   2 if there was just the right amount of training on this topic
   3 if there was too much training on this topic

(NOTE: If you believe there was too much or too little training in any of the topic areas listed below, please specify in the Comments section below the subject matter, e.g., statutes, ACS procedures, screen displays, etc., and describe your recommended change.)

	Rating
Training in Collection/ACS procedures	....
Training in Tax Law	....
Training in skiptracing for investigation	....
Training in forms preparation	....
Training in interview/telephone collection techniques	....
Other (specify	
....	

Comments: ____

____

____

105. The most frustrating part of getting your work done is:

....

....

....

106. The most rewarding part of your work is:

....

....

....

107. If you could "realistically" change one part of your job, it would be:

....

....

....

108. If you could keep only one part of your job, it would be:

....

....

....

109. If you were able to change your position by adding or removing certain of your current responsibilities, what type of changes would you recommend?

....

....

....

Appendix 3
Recruitment and Retention Instruments

Recruitment/Attrition Survey

As an employee of Collection, you are being asked to take part in a study looking into the reasons why people leave and why they stay in the Automated Collection System. Your participation is optional. If you chose to participate, the opinions you express on this survey will be taken as representative of many other ACS employees who will not have the opportunity to be a part of this study.

Please do not place your name on any part of this survey. All answers you make will be combined with others. Individual responses will not be identified.

This survey will result in a series of recommendations to make ACS a more desirable and productive place in which to work. Your opinion will have an effect on these recommendations. NTEU will be notified and provided the opportunity to bargain prior to any recommendations being implemented. Thank you for your participation.

Department of the Treasury
Internal Revenue Service
Form M-7191 (10-85)

I Background Data

1. Current Age (check one):

☐ less than 20 ☐ 36 to 40 ☐ 56 to 60
☐ 21 to 25 ☐ 41 to 45 ☐ 61 to 65
☐ 26 to 30 ☐ 46 to 50 ☐ 66 or older
☐ 31 to 35 ☐ 51 to 55

2. Sex:

☐ male ☐ female

3. Marital Status (check appropriate category):

☐ single, never married ☐ married
☐ divorced/widowed ☐ other

4. Highest educational level:

☐ no high school degree ☐ bachelor's degree
☐ high school/GED degree ☐ some graduate courses
☐ master's degree
☐ some college ☐ doctoral degree
☐ associate degree

5. How long have you been with ACS?

_____ years _____ months

6. How long have you been with the IRS?

_____ years _____ months

7. How did you first hear of the job in ACS? (check one):

☐ newspaper advertisement ☐ fellow employee
☐ friend or relative in the IRS
☐ Civil Service (OPM) ☐ other source (please specify) _______________
☐ radio/television
☐ IRS vacancy notice

8. What is your job series and grade? __________ Job series
__________ Grade

9. How long have you been at your current grade level?

_____ years _____ months

10. Employment status:

☐ permanent ☐ WAE/seasonal
☐ temporary

11. How many people depend on your IRS income? (including yourself)

a) Number who are totally dependent on you: _____
b) Number who are partially dependent on you: _____

12. What percent of your total family income from all sources comes from your IRS salary? _______ %

13. Within the last three years, how many times have you changed jobs within the IRS (i.e., had different position descriptions): ________

14. Within the last five years, how many other jobs have you held in the private sector or other government organizations (federal, state, county, or local)? ________

15. In terms of what is required of you, how different is your current ACS job compared to prior non-ACS jobs?

☐ very different ☐ similar
☐ different ☐ very similar

16. To what extent do you use your educational background or training in your current job?

☐ a very great deal ☐ a little
☐ a great deal ☐ very little
☐ somewhat

17. How many miles (round trip) do you commute to work?

___ miles round trip

18. How do you commute to work?

☐ alone in my car ☐ by bus and/or train
☐ in a car or van pool ☐ other means (walk, ride a bike, etc.)

19. On the average, how long does it take you to commute to work round trip?

_____ hrs. _____ minutes

II Work/Job Preference

Of the factors listed below place an "M" next to those five (5) items which you are MOST important to you in a job, then place an "L" next to those five factors which are LEAST important to you.

(1) ☐ work for a small organization
(2) ☐ work at a relaxed pace
(3) ☐ work in a competitive, challenging environment
(4) ☐ work near where I live
(5) ☐ work at a high risk/high reward job
(6) ☐ work for a supervisor I respect
(7) ☐ work at a job that has a set routine and structure
(8) ☐ work at a low-stress job
(9) ☐ work at a complex job
(10) ☐ work which involves contact with the public
(11) ☐ work in an informal, people oriented environment
(12) ☐ work in a high prestige job
(13) ☐ work that is technical in nature
(14) ☐ work that involves supervising others
(15) ☐ work that allows you to operate independently of others
(16) ☐ work that involves considerable job security
(17) ☐ work that gives you a chance to learn new skills
(18) ☐ work that provides numerous opportunities for advancement
(19) ☐ work that pays a good wage
(20) ☐ work that allows outside employment

(21) ☐ work in which you make a real difference in product quality
(22) ☐ work that has a real feeling of accomplishment
(23) ☐ work that involves good co-workers
(24) ☐ work in an office with pleasant physical surroundings
(25) ☐ work that involves a measurable output
(26) ☐ work in which I get regular feedback on my performance
(27) ☐ work in which I have control over my work activities
(28) ☐ workloads that are predictable and evenly paced
(29) ☐ work with the right equipment and procedures to get the job done
(30) ☐ work in which the employees are treated fairly by management

8. If you could do it all over, how could you have improved your career with the IRS?

____
____
____
____
____
____
____

III Job Satisfaction

1. How do you feel about your present job?
   ☐ very satisfied ☐ dissatisfied
   ☐ satisfied ☐ very dissatisfied

2. How do you feel about your overall career progress to date?
   ☐ very satisfied ☐ dissatisfied
   ☐ satisfied ☐ very dissatisfied

3. How do you feel about your future career prospects with the IRS?
   ☐ very satisfied ☐ dissatisfied
   ☐ satisfied ☐ very dissatisfied

4. How important to you is work compared to the other things in your life?
   ☐ very important ☐ unimportant
   ☐ important ☐ very unimportant

5. How would *you* rate your current level of effectiveness as an employee compared to other ACS employees?
   ☐ far superior to others ☐ slightly less effective
   ☐ superior to others ☐ a lot less effective
   ☐ about the same as others

6. How would your *supervisor* rate your current level of effectiveness compared to other ACS employees?
   ☐ far superior to others ☐ slightly less effective
   ☐ superior to others ☐ a lot less effective
   ☐ about the same as others

7. Would you recommend the IRS as a career to others just starting out?
   ☐ absolutely yes ☐ a qualified no
   ☐ a qualified yes ☐ absolutely not

IV Attrition

1. What is the chance that in the next year you will leave ACS for another position *within* the IRS?
   ☐ 100% chance ☐ 25% chance
   ☐ 75% chance ☐ 0% chance
   ☐ 50% chance

2. What is the chance that in the next year you will leave ACS for another position outside of the IRS?
   ☐ 100% chance ☐ 25% chance
   ☐ 75% chance ☐ 0% chance
   ☐ 50% chance

3. Check *one* of the following which best tells how you feel about changing your job:
   ☐ I would quit this job at once if I could get anything else to do
   ☐ I would take almost any other job where I could earn the same salary
   ☐ I would like to change both my job and my occupation
   ☐ I am not eager to change my job but I would do so for a better job
   ☐ I cannot think of any jobs I would exchange for mine

4. If you were to leave your job, it would be for the following reasons (Check all that apply):
   ☐ education, return to school
   ☐ desire for permanent employment
   ☐ desire for full time employment
   ☐ more pay/promotion
   ☐ future career opportunities
   ☐ more interesting/challenging work
   ☐ poor co-worker relations
   ☐ poor management practices/policies
   ☐ poor physical working conditions
   ☐ poor work performance
   ☐ personal reasons (transportation, health, moving, home responsibilities)
   ☐ production pressure/job stress
   ☐ retirement
   ☐ other: (specify)

   ____
   ____
   ____

5. If you were to leave your job, where would you most likely go (check one)?
   ☐ promotion within my job series
   ☐ promotional transfer to another IRS office
   ☐ lateral transfer to another IRS office
   ☐ another federal agency
   ☐ a state agency

ACS Resignation Notice

(Please fill in the following information and mail immediately)

Name: ______________________________
(Last) (First) (M.I.)

Left ACS on: ____/____/____

Call Site: ______________________________

Form **M-7193** (10-85) Department of the Treasury—Internal Revenue Service

ACS Exit Card

Why did you leave ACS? ______________________________

Where did you go after leaving ACS? ______________________________

(Please mail this card immediately after completing.)

Form **M-7194** (10-85) Department of the Treasury—Internal Revenue Service

References

Etzioni, A., 1961, *A Comparative Analysis of Complex Organizations* (New York: Free Press of Glencoe).

Herzberg, F., Mausner, B. and Snyderman, B., 1959, *The Motivation to Work* (New York: John Wiley & Sons).

Hirschein, R. A., 1985, *Office Automation* (Reading, MA: Addison-Wesley), pp. 224–237.

Howard, J. L. 1983, Overview of office automation. Speech presented before the Federal Office Systems Exposition, 1983, Washington, DC.

Katzell, R. and Yankelovich, D., 1975, *Work, Productivity and Job Satisfaction* (Washington, DC: The Psychological Corp.).

Lewin, K., 1947, *The Research Center for Group Dynamics.* Sociometry Monograph No. 17 (New York: Beadon House).

Marguiles, N. and Colfesh, A., 1982, A socio-technical approach to the planning and implementation of new technology. *Training and Development Journal,* Dec.

Maslow, A. H., 1970, *Motivation and Personality* (New York: Harper and Row).

Mobley, W. H., 1982, *Employee Turnover: Causes, Consequences, and Control* (Reading, MA: Addison-Wesley).

Parsons, T., 1960, Some ingredients of a general theory of formal organizations. In *Organizations: Systems, Control and Adaption*, edited by J. A. Litterer (New York: John Wiley & Sons), pp. 197–213.

Stettner, L., 1966, Survey of literature on social and economic effects of technical change. In *Employment Problems of Automation and Advanced Technology: An International Perspective* (London: Macmillan).

Strauss, G., 1977, Managerial practices. In *Improving Life at Work*, edited by J. R. Hackman and J. S. Suttle (Santa Monica, CA: Goodyear Publishing), pp. 297–363.

Taylor, F. W., 1947, *Principles of Scientific Management* (New York: Harper and Row).

Walton, R. E., 1980 Establishing and maintaining high committment work systems. In *The Organizational Life Cycle* (San Francisco: Jossey-Bass).

Zuboff, S., 1982, New worlds of computer-mediated work. Harvard Business Review, **60** (5), 142–152.

Additional Resources

Bailey, R. W., 1982, *Human Performance Engineering: A Guide for Systems Designers* (Englewood Cliffs, NJ: Prentice Hall).

Brill, M., 1981, *The Impact of Office Environment on Productivity and Quality of Working Life: Comparative Findings* (Buffalo, NY: The Buffalo Organization for Sociological and Technological Innovation).

Card, S. K., Moran, T. P. and Newell, A., 1983, *The Psychology of Human-Computer Interaction* (Hillsdale, NJ: Lawrence Erlbaum).

Cohen, E. and Cohen, A., 1983, *Planning the Electronic Office* (New York: McGraw-Hill).

Datapro Research Corp., 1985, *Managing Office Automation,* Vols. 1 and 2 (Delran, NJ: Datapro).

Educational Services, Digital Equipment Corp., 1983, *Office Automation Implementation* (Cambridge, MA: Digital Equipment Corp.).

Gery, G., 1982, Office technology: creating receptivity among executives and professionals. *National Productivity Review,* **1** (2).

Grandjean, E., 1980, *Fitting the Task to the Man* (London: Taylor & Francis).

Johnson, B. M., 1985, *Innovation in Office Systems Implementation.* National Science Foundation Report No. 8110791 (Washington, DC: Government Printing Office).

Lawler, E. E., Nadler, D. and Cammann, C., 1980, *Organizational Assessment* (New York: John Wiley & Sons).

Ledin, H., 1983, *Ergonomic Principles in Office Automation* (Stockholm: Ericsson Information Systems AB).

Leonard-Barton, D., 1985, Implementing new technology. *Harvard Business Review,* **63** (6).

National Science Foundation, 1983, *The Process of Technological Innovation: Reviewing the Literature* (Washington, DC: Productivity Improvement Research Section, Division of Industrial Science and Technology Innovation).

Office of Technology Assessment, 1985, *Automation of America's Offices, 1985–2000.* Document No. OTA-CIT-287 (Washington, DC: US Government Printing Office).

Rubin, A., 1983, *The Automated Office - An Environment for Productive Work or an Information Factory?* (Washington, DC: Center for Building Technology, National Bureau of Standards).

Seashore, S. E. *et al.*, 1983. *Assessing Organizational Change* (New York: John Wiley & Sons).

Sell, R. G. and Shipley, P., 1979, *Satisfactions in Work Design: Ergonomics and Other Approaches* (London: Taylor & Francis).

Tichy, N. M., 1983, *Managing Strategic Change* (New York: John Wiley & Sons).

Uhlig, R. P., 1979, *The Office of the Future: Communication and Computers* (Amsterdam: North Holland).

US Dept of the Treasury, Internal Revenue Service, 1984, *Human Factors in the Implementation of the Automated Collection System.* Document No. 6985 (Washington, DC: Government Printing Office).

US Dept of the Treasury, Internal Revenue Service, 1985, *A Guidebook: How to Introduce Ergonomics in IRS Automation Projects.* Document No. 7010 (Washington, DC: Government Printing Office).

Van de Ven, A. and Ferry, D., 1980, *Measuring and Assessing Organizations* (New York: John Wiley & Sons).

Westin, A. F. *et al.* 1985, *The Changing Workplace: A Guide to Managing the People, Organizational, and Regulatory Aspects of Office Technology* (White Plains, NY: Knowledge Industry Publications).

Index

For Product Safety Concerns and Information please contact our EU
representative GPSR@taylorandfrancis.com
Taylor & Francis Verlag GmbH, Kaufingerstraße 24, 80331 München, Germany

www.ingramcontent.com/pod-product-compliance
Lightning Source LLC
Chambersburg PA
CBHW071351220926
51613CB00008B/47

9780415823265